99 Maps to Save the Planet

CREATED AND DESIGNED BY KATAPULT MAGAZINE

Text translated from the German by Jefferson Chase

With an Introduction by Chris Packham

THE BODLEY HEAD
LONDON

1 3 5 7 9 10 8 6 4 2

The Bodley Head, an imprint of Vintage,
One Embassy Gardens
8 Viaduct Gardens
London
SW11 7BW

The Bodley Head is part of the Penguin Random House group of companies whose addresses can be found at global.penguinrandomhouse.com.

Copyright © Suhrkamp Verlag 2020
Translation copyright © Jefferson Chase 2021
Introduction copyright © Chris Packham 2021

Katapult Magazine has asserted its right to be identified as the author of this Work in accordance with the Copyright, Designs and Patents Act 1988

Jefferson Chase has asserted his right to be identified as the translator of this Work in accordance with the Copyright, Designs and Patents Act 1988

First published as *102 Grüne Karten zur Rettung der Welt* by Suhrkamp Verlag in 2020

First published by The Bodley Head in 2021

www.penguin.co.uk/vintage

A CIP catalogue record for this book is available from the British Library

ISBN 9781847926500

Typeset in 10/14 pt Frutiger LT Com
by Integra Software Services Pvt. Ltd, Pondicherry

Printed and bound in Latvia by Livonia Print

The authorised representative in the EEA is Penguin Random House Ireland, Morrison Chambers, 32 Nassau Street, Dublin D02 YH68.

This book is printed on 100% recycled paper.

Penguin Random House is committed to a sustainable future for our business, our readers and our planet. This book is made from Forest Stewardship Council® certified paper.

This printed matter is carbon compensated according to ClimateCalc.
Offsets purchased from: South Pole
www.climatecalc.eu
Cert. no. CC-000090/LV

Introduction

Terrifying yet funny, surprising yet predictable, simple yet poignant, occasionally abstract and yet so very, very important – this collection of infographics presents a telling picture of our planet at the beginning of the 21st century. There is both a ruthless clarity and beautiful precision in the design. These maps are unforgiving: once seen they are tattooed into your memory because they distil the horror of our careless and destructive inadequacies so explicitly. And thus they become not just revealing but pertinently motivating too. These pictures are tools – power tools in a kit that must be employed as rapidly as possible to rectify the grotesque realities they so brutally expose. They tell the truth: a truth we must confront to enable change; indeed, they have been conceived to change our minds and they leave us with no doubt that without success in this regard we are doomed.

These maps work brilliantly because the premise behind each message is so straightforward, devoid of clutter or circuitous pernickety explanation; because the visual carrier of that message is so bold and clear, at times wonderfully childlike; and because the use of regional and global maps instantly identifies our place in this catastrophe and connects us with everyone else. Cleverly the authors include breathers too – nuggets of information which are just nice to know, some silly humour, seemingly trivial facts to rest our conscience – before we are shaken awake again by the terrible truth told by another piece of smartly presented data.

We are a visual species, we can communicate in a universal language of signs, symbols and gestures. We distil immeasurably complex emotions into a smile, we stop at the depiction of a cross and we part with the wave of a hand. And as much as we embrace complexity and revel in all its nooks and crannies, we can also be bedevilled by detail. Perhaps we have got lost in the existential labyrinth of memes and messages that we've generated around climate change and biodiversity: perhaps we've made it too difficult to understand, interrelate and imagine. Here is the antidote. There is no struggle to grasp the facts displayed so lucidly here, no excuse to ignore them.

These graphics tell our story, a story which is begging for a happier ending . . .

Chris Packham

Have the courage to use your own reason

This book doesn't provide any action plan to protect our climate and our planet; there are no practical tips, no top ten of the best green activities. KATAPULT avoids that sort of thing: we only present the facts. Why? In the eighteenth century a famous philosopher told his students: 'Think for yourself and then act on it.' He also produced this punchy one-liner: 'Have the courage to use your own reason!' People liked that sentence so much it became the motto of the Enlightenment, the revolution in ideas that created the modern world.

We believe our readers don't need to be told what to do. If you are shocked by the facts presented in this book, then you will want to do something about it. And you will have your own ideas, big or small, about what you can do to protect the environment. Every individual can make things better. But that's not enough: the decisive change has to be political. And for that we need a different breed of politicians – not those who short-sightedly play climate protection off against jobs and economic prosperity or who postpone plans for climate protection until 2050 because they can't be bothered.

We started KATAPULT magazine in 2016 to make science accessible and exciting. We use infographics and maps to question boring traditions and lazy assumptions and to deliver fresh ideas and arguments. And as you can see, we now also put together books.

This book reveals the precarious state of our planet – but we also hope it shows how easy it would be to improve things.

Benjamin Fredrich
Editor-in-Chief
KATAPULT

Active coal-burning power plants

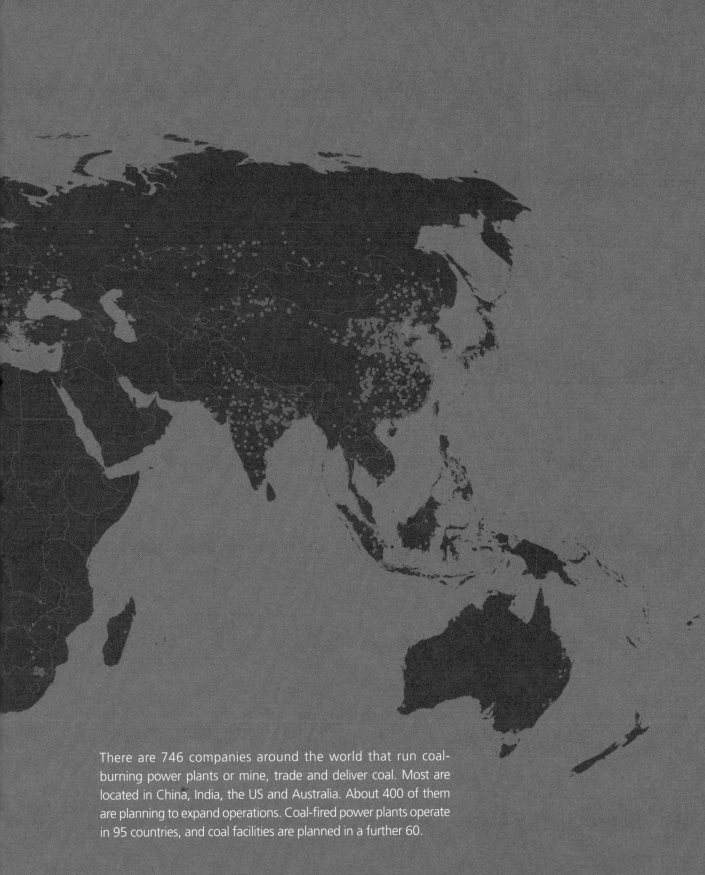

There are 746 companies around the world that run coal-burning power plants or mine, trade and deliver coal. Most are located in China, India, the US and Australia. About 400 of them are planning to expand operations. Coal-fired power plants operate in 95 countries, and coal facilities are planned in a further 60.

How We see nature

How Nature sees us

Shark vs man in 2018

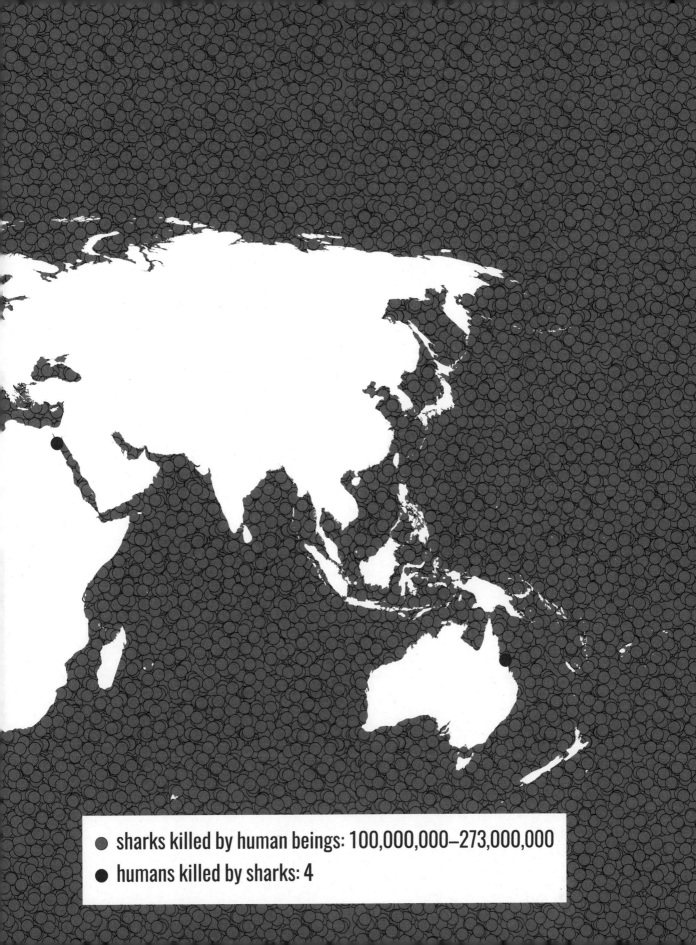

Wolves in Europe

By the mid-nineteenth century, the wolf was considered extinct in many parts of northern and central Europe. In Britain, that's still the case despite efforts to reintroduce the animal. But over the past decades, wolves have returned from the south and the east to some areas in central Europe. They re-emerged in Germany, for instance, in 2000 and have been allowed to live there; they are even protected by law.

Still, wolves are unlikely to spread back to the entire continent. In many regions of Europe too many roads cut through the environment and there isn't enough wildlife to support them. By the way, statistically speaking, people in Western countries are far more likely to be attacked by a dog than by a wolf.

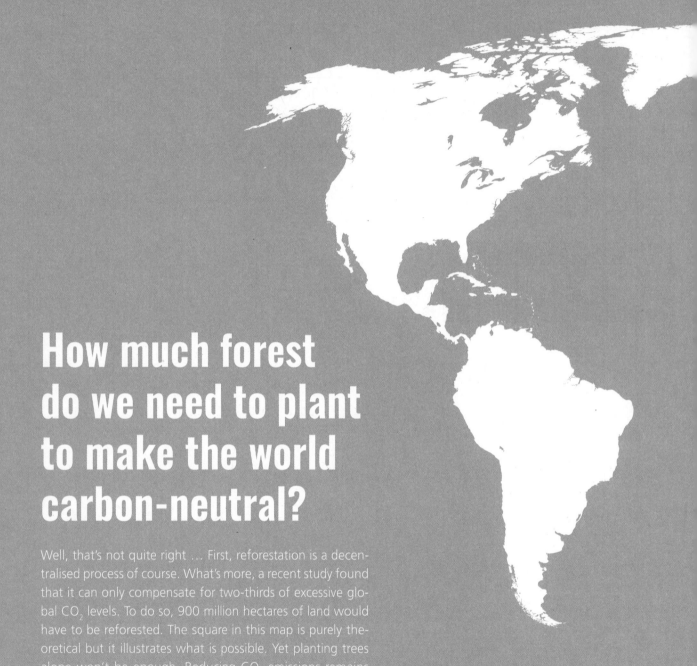

How much forest do we need to plant to make the world carbon-neutral?

Well, that's not quite right … First, reforestation is a decentralised process of course. What's more, a recent study found that it can only compensate for two-thirds of excessive global CO_2 levels. To do so, 900 million hectares of land would have to be reforested. The square in this map is purely theoretical but it illustrates what is possible. Yet planting trees alone won't be enough. Reducing CO_2 emissions remains our top priority if we want to protect our climate.

Reforestation is possible in the green areas

In the other regions, no new forests can be planted either because they're already covered with trees or because the ground is too arid or frozen to support them. Reforestation done the wrong way can also be damaging. China has resorted to monoculture and planted large areas with only one type of tree. This has caused excessive acidity in the soil – an environmental catastrophe.

Every third piece of rubbish in the sea is a cigarette butt

Every year 4.5 trillion cigarette butts end up in the environment. One such remnant is enough to contaminate around 40 litres of water, and cigarette ends account for every third bit of rubbish that washes up on shore. Fish ingest toxic substances and microplastics before they end up being served in restaurants.

Shrinkage of Switzerland's major glaciers between 1850 and 2010

Roughly half of the area of Switzerland's glaciers has melted away since 1850. As the gigantic masses of snow and ice disappear, so too do the plants and animals that lived there. The cause is lack of water. At the same time, the ground has become unstable and landslides are more likely.

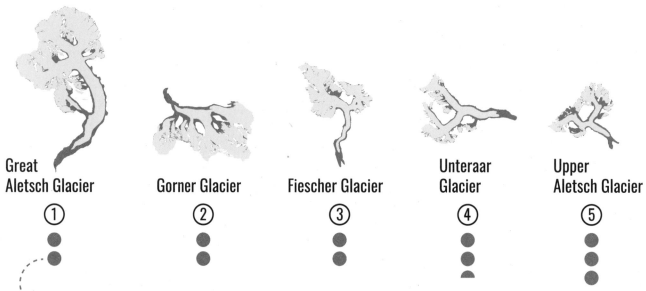

Shrinkage in per cent: every blue dot indicates 10 per cent (data for 1850–2016)

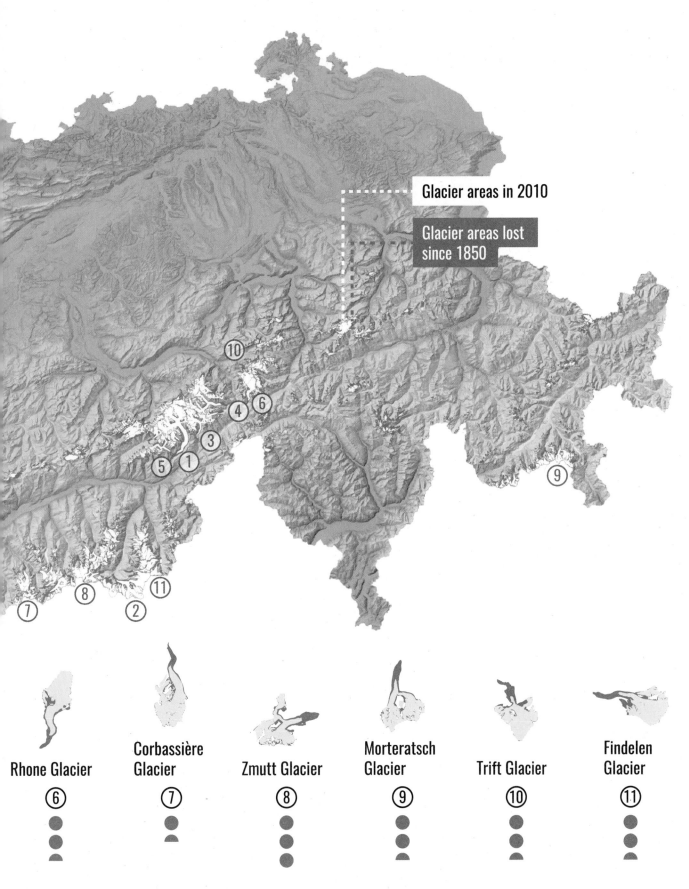

Things that are getting too warm

Things we object to being too warm

If all the people in the world stood next to one another, they'd cover an area of 62 x 62 km

The earth is overpopulated – this common assertion suggests that the planet doesn't have enough resources for all of its 7.6 billion inhabitants. But what about the space they occupy? If two people shared one square metre, the earth's entire population would cover 3,844 square kilometres. Texas alone is 176 times bigger than that.

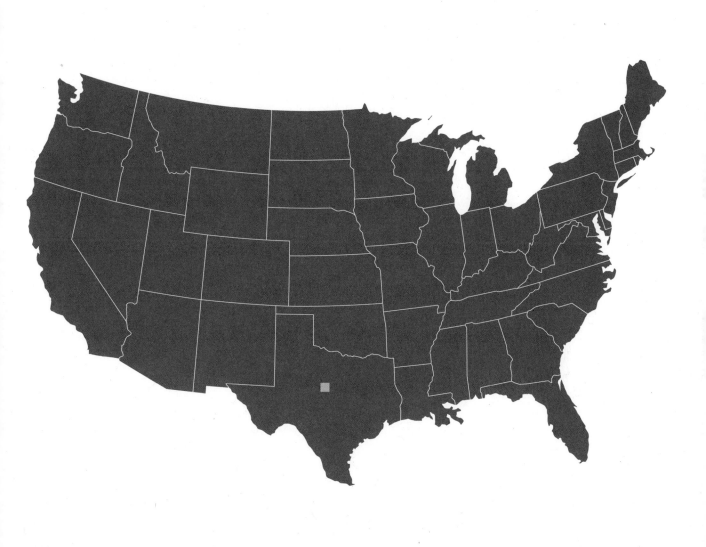

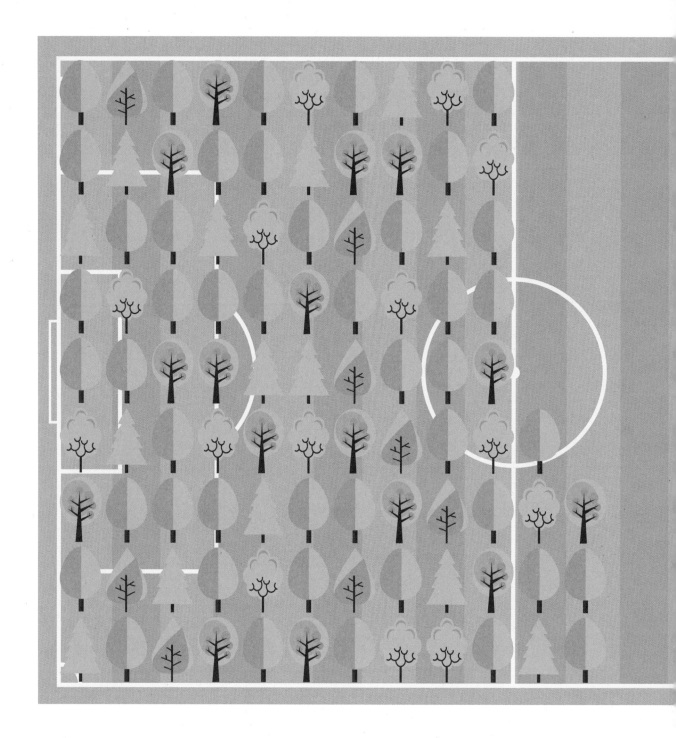

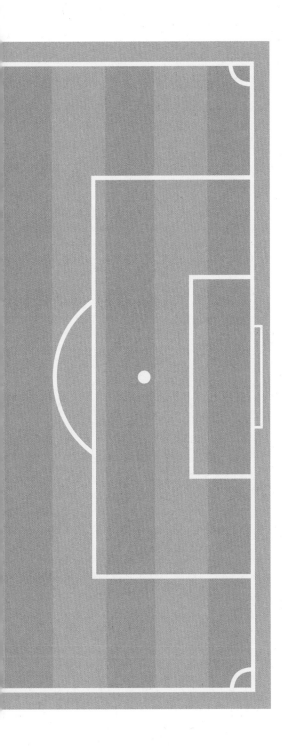

Since human beings became sedentary 12,000 years ago, the number of trees worldwide has declined by 46 per cent.

This is what this deforestation would look like if the world were a football pitch.

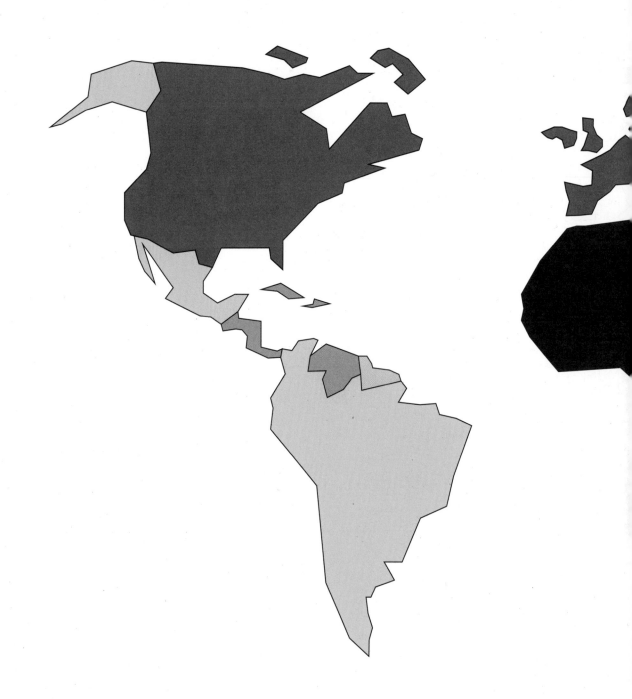

How Westerners react to catastrophes in these parts of the world

3,700 km

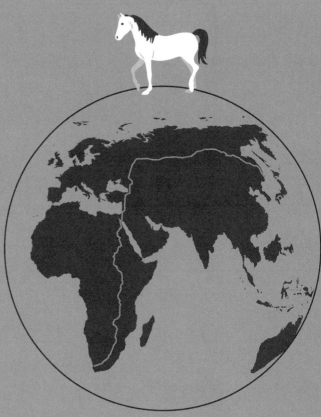

23,500 km

Keeping a horse has the same impact on the environment as a 23,500-kilometre road trip

The larger and heavier our pets, the more of a burden they are on our environment. Animals need to be fed, housed, transported, etc. This graphic represents not only the CO_2 figures but shows so-called environmental impact points, which alongside carbon emissions also include the use of resources. Expressed in terms of pets: keeping one cat is the equivalent of keeping two rabbits, ten birds or around 125 fish. It's important to note that these figures aren't averages but model calculations. By the way, the average annual mileage of a car in the UK is 12,231 kilometres.

1,500 km

The Great Green Wall

Africa's Great Green Wall is one of the largest environmental projects in the world. Inaugurated in 2005 by the African Union and aimed at combatting desertification and other effects of climate change in the Sahel, the initiative envisioned creating a 15-kilometre-wide corridor of trees stretching between the west and east coasts of the planet's largest continent. But the wall has turned into a mosaic, after the original idea proved neither ecologically nor socially feasible. In some regions it is impossible to cultivate trees because of desiccation; in others, the planting of trees would have disrupted local agriculture. As a result, the initiative now supports decentralised, local projects. More than twenty nations have taken part in the scheme and twelve million trees have already been planted.

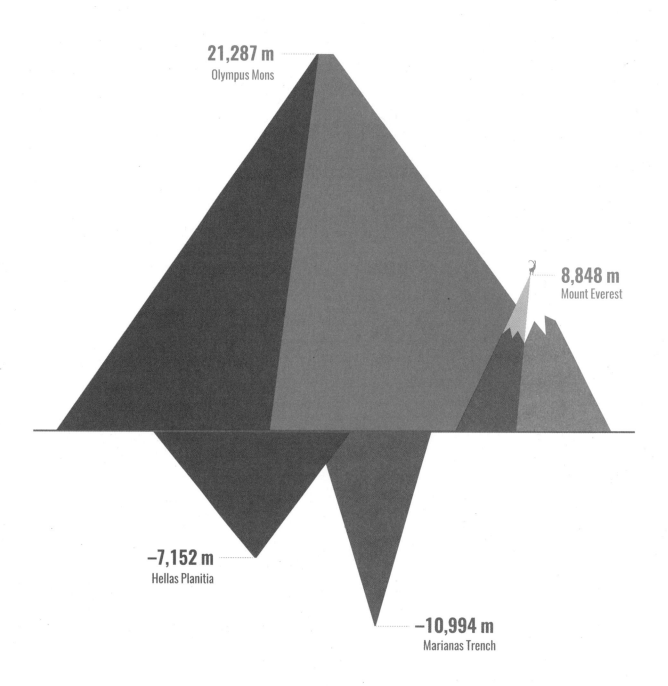

The highest and lowest points on Earth and Mars

The mass of Earth is ten times that of our neighbour Mars. Nonetheless, the much smaller Red Planet has a mountain that is two-and-a-half times the size of Mount Everest. The peak of the Martian volcano Olympus Mons is 21,000 metres above the middle plateau region – unlike on Earth, heights on Mars aren't measured with reference to sea level. In fact, Olympus Mons is the tallest planetary mountain in the entire solar system. It's topped only by a formation called Rheasilvia on the asteroid Vesta between Mars and Jupiter – its cratered peak is more than 22 kilometres tall.

$ Countries that index GNP

♥ Countries that index people's wellbeing

To hell with GNP – that's the attitude of Iceland, Scotland and New Zealand. Since 2018, the political leaderships of those three countries have designated themselves 'Wellbeing Economy Governments'

that prioritise the happiness of their populations over economic growth. A radically new idea? Not really. The constitution of Bolivia, for example, has enshrined the fair distribution of natural resources as an official goal of state. The French government uses a Social Justice Index. And since 1997, Bhutan has officially measured 'gross national happiness'.

Numbers of cruise ship passengers keep going up

Cruises may be the dirtiest form of travel on earth, but that doesn't deter the many people who holiday at sea. The average cruise passenger spends nine days on board ship, with the most popular destinations being northern Europe, the Baltic Sea, the Mediterranean, the Canary Islands and the Caribbean. Most of these tourists fly to and from their cruise holidays and thereby further increase their carbon footprint.

3.8

Numbers of cruise ship passengers worldwide in millions

4.7

7.2

1995

2000

26

22.5

18.4

	Titanic	**Symphony of the Seas**
Displacement:	c.50,000 t	c.100,000 t
Length:	269.1 m	362.1 m
Width:	28.3 m	65.7 m
Passengers:	2,603	5,518

1.2

Symphony of the Seas

Titanic

2005　　　　　　　　　2010　　　　　　　　　2015

Annual expenditure per person for cycle paths in Europe

In the centre of the Dutch city of Utrecht, 60 per cent of all traffic consists of cyclists. Why is that? The city not only builds many cycle paths but also has the largest multi-storey cycle park in the world. Cyclists look for safety, so that's why a cycle path that is separated from the road is important. In many other countries, cycle paths are on the road so that cyclists and cars move in close proximity. As a result many cyclists feel unsafe and stop using their bikes.

BERLIN — 4,70 €

AMSTERDAM — 11 €

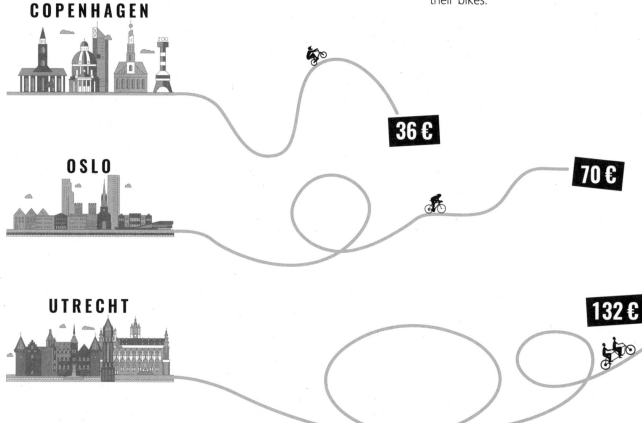

COPENHAGEN — 36 €

OSLO — 70 €

UTRECHT — 132 €

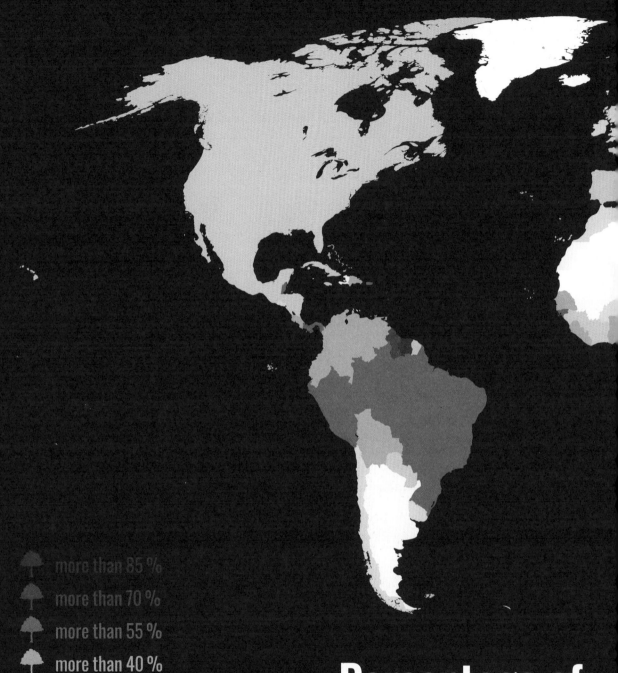

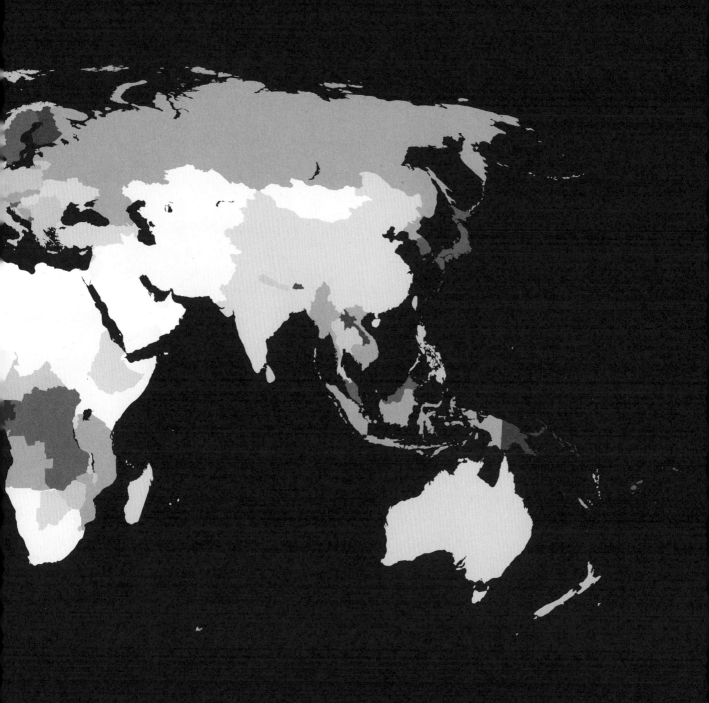

At the moment, there are 400 trees for every human being on this planet, but that ratio is decreasing. Generally speaking, as the world's population grows, forests disappear, but there are also significant local variations around the globe. Between 1990 and 2015, forested areas in affluent countries grew on average by 1.3 per cent every year, whereas in poor countries they decreased by 0.7 per cent.

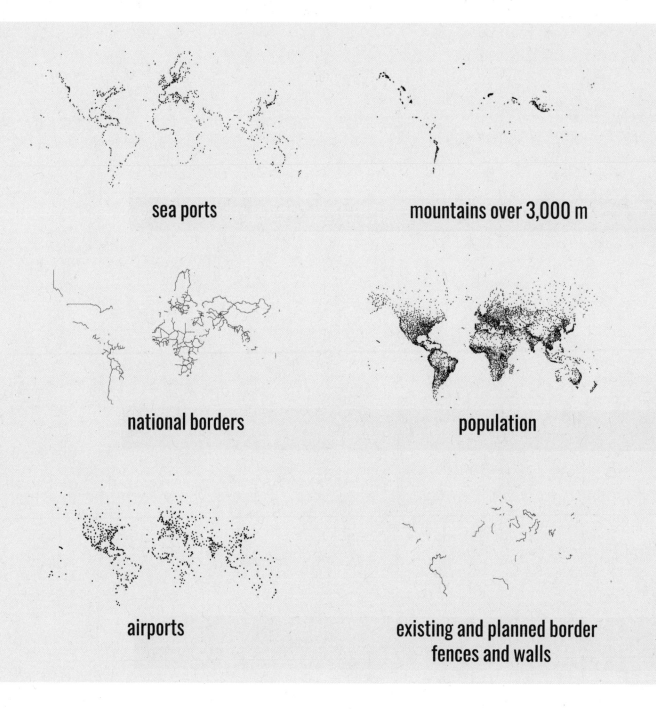

Minimalist maps

We can recognise our world on maps even when they don't take the form we are used to. The basic grid of a world map is usually provided by land masses and national borders. If these are omitted, and maps instead depict human artefacts, we can still see familiar outlines – but from a new perspective. Sea ports are

underseas cables

active nuclear power plants

90 most widely read newspapers

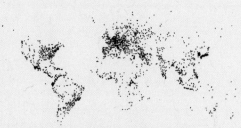
universities

roads

national capitals

located solely on the edges of continents. Europe has a comparatively large number of universities. Sparsely populated and uninhabited areas like the Sahara or eastern Russia disappear from the map entirely.

Particulate matter (PM) and air quality in 2017

People in Montenegro tell the following joke. A resident of Pljevlja, one of the country's most polluted cities, travels to Lovćen National Park. When he gets off the bus, he passes out. 'Quick!' shouts the bus driver. 'Put him under the exhaust so he can breathe again!'

The point of the joke applies to all Balkan countries, but perhaps most to Bosnia-Hercegovina. According to a World Health Organization report in 2017, that nation has the highest levels of particulate matter in Europe and leads the statistics for air pollution fatalities.

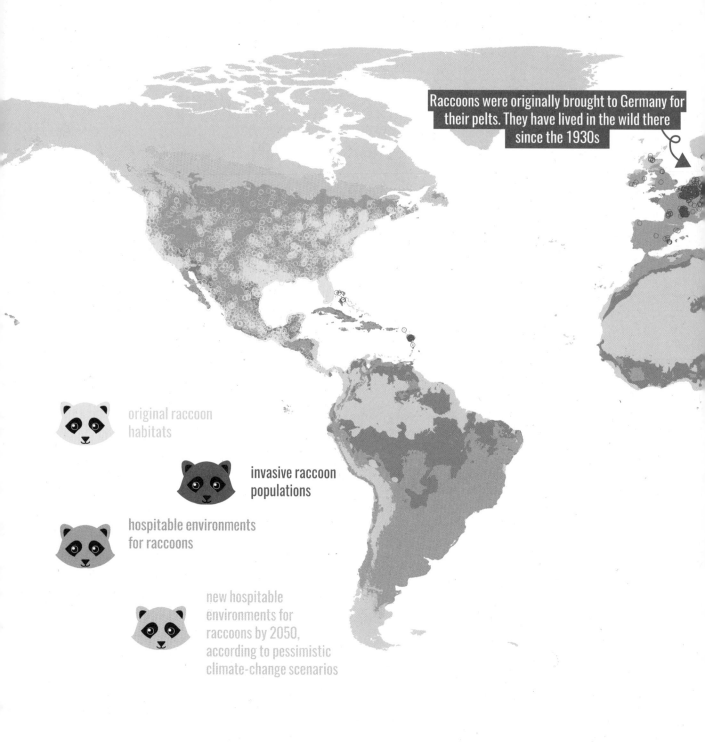

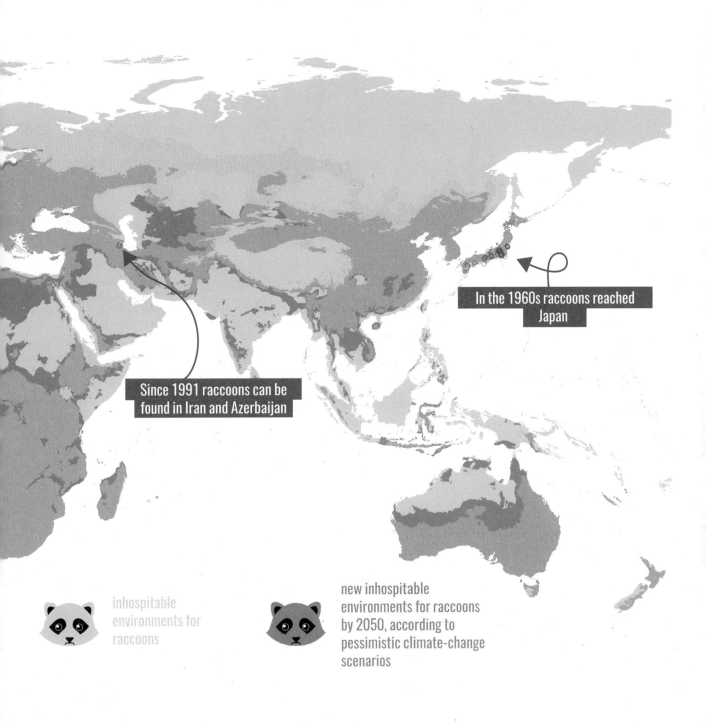

Originally native only to North America, raccoons were introduced in other countries for their pelts. There they soon spread to the wild. In places like Japan, Iran and Azerbaijan, the omnivorous creatures endanger the habitats of native animals, agriculture and the entire ecosystem.

Minimalist map of Saudi Arabia, showing its rivers

Saudi Arabia has neither rivers nor lakes, merely a few wadis – valleys and dry riverbeds that only conduct water after strong rainfall. To avoid having to import water, the Saudis have to dig wells deep into the earth's surface and rely on desalination facilities. But water procured from the sea not only has to have the salt removed. It also has to be thoroughly decontaminated since oil production has badly polluted the Red Sea and Persian Gulf coasts.

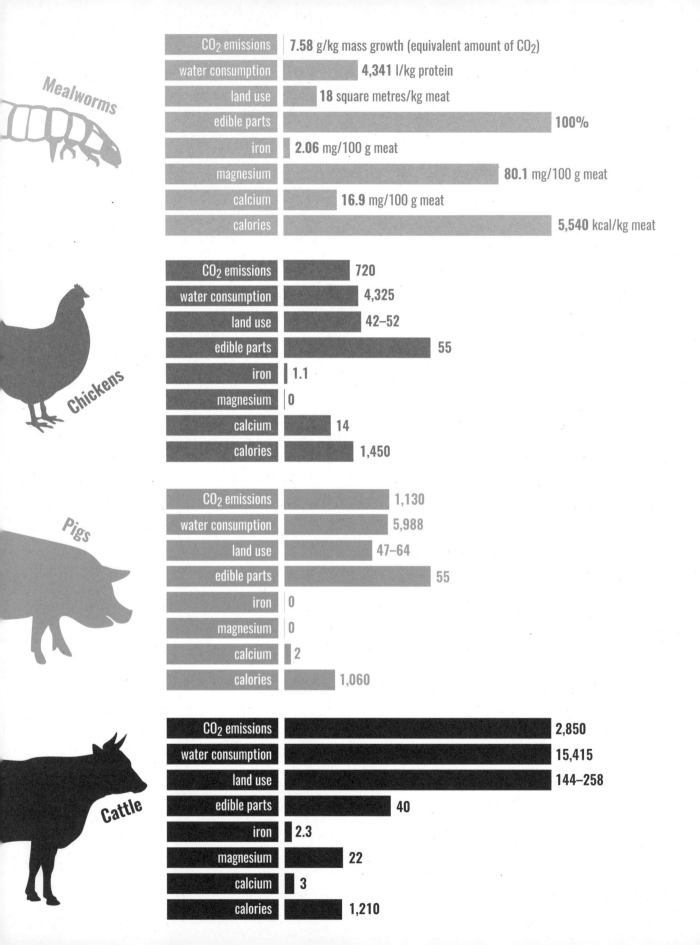

Consumption of resources vs nutritional yield of various animals

Three people go to a restaurant. One orders a meat dish, another something vegetarian and the third a burger made of mealworms. All of them contain between 915 and 940 kilocalories, and each of the diners feels full after the meal. The difference is that the person who eats the mealworm burger caused the least CO_2 emissions – just 160 grams. The vegetarian is responsible for 470 grams, and the meat-eater for 2,020 grams. In many regions of the world, particularly in Asia, people eat insects, of which there are more than 2,100 edible varieties.

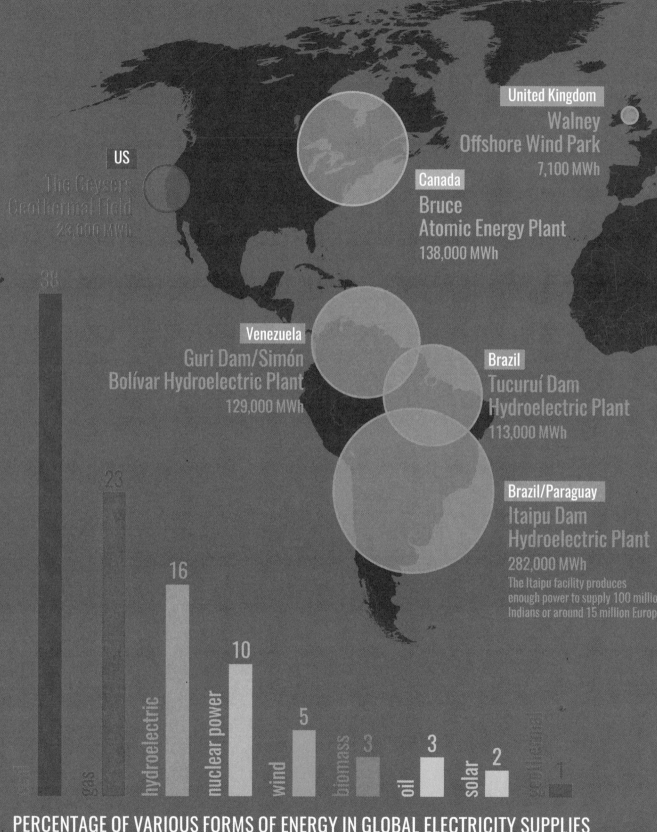

PERCENTAGE OF VARIOUS FORMS OF ENERGY IN GLOBAL ELECTRICITY SUPPLIES

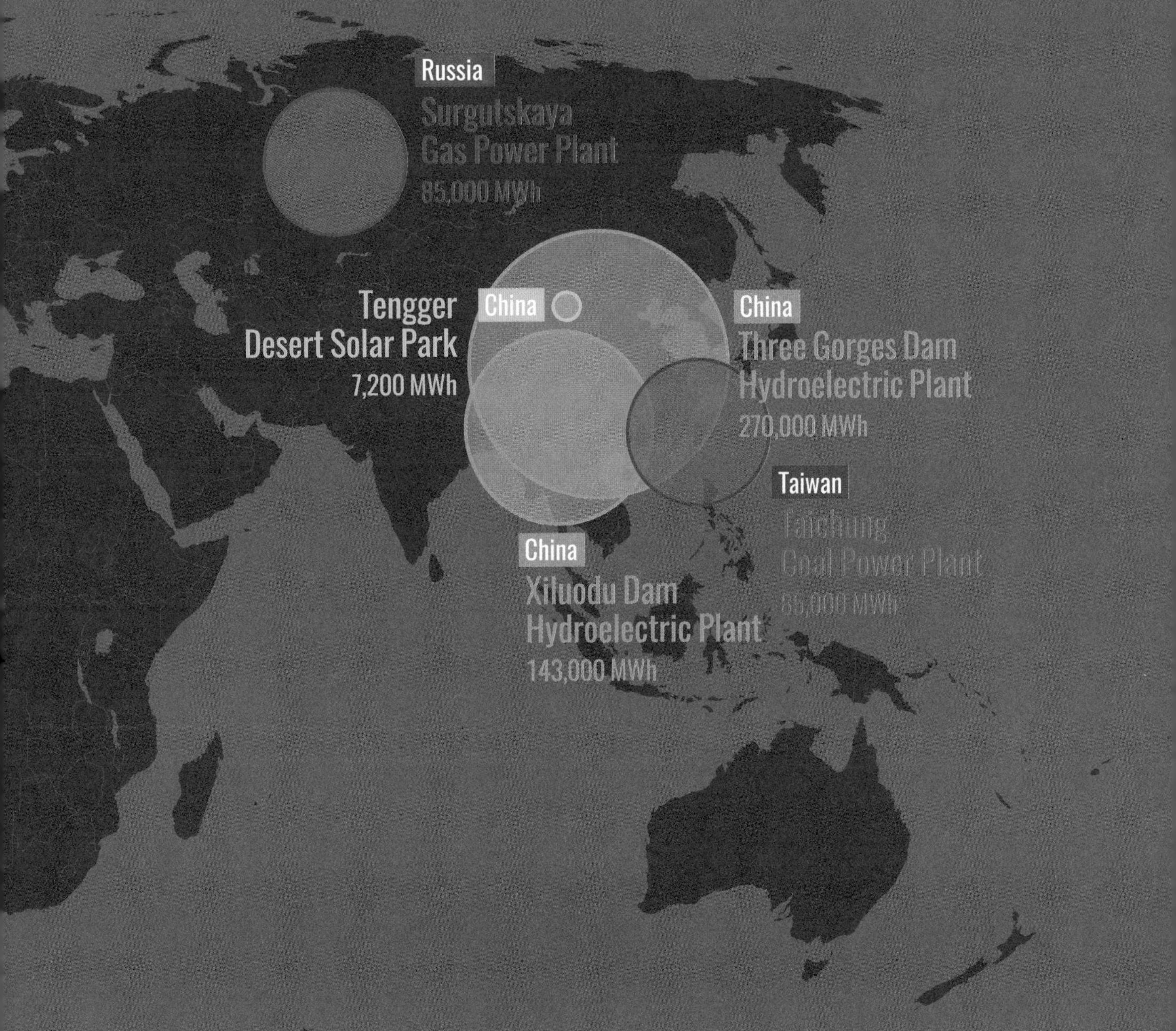

Russia
Surgutskaya
Gas Power Plant
85,000 MWh

China
Tengger
Desert Solar Park
7,200 MWh

China

China
Three Gorges Dam
Hydroelectric Plant
270,000 MWh

China
Xiluodu Dam
Hydroelectric Plant
143,000 MWh

Taiwan
Taichung
Coal Power Plant
85,000 MWh

The five largest hydroelectric facilities compared to the most productive power plants for other energy types (in electricity generated per day)

Nations that used a higher or lower percentage of renewable energies than the UK in 2018

Almost all the countries of the EU have made more progress towards transitioning to renewable energy sources than the United Kingdom. Some of these nations have geographical advantages. Southern countries have more sunshine, while hydroelectric power can be generated in mountainous regions. Britain has neither major mountain ranges nor lots of sun so it has focused on developing offshore wind energy.

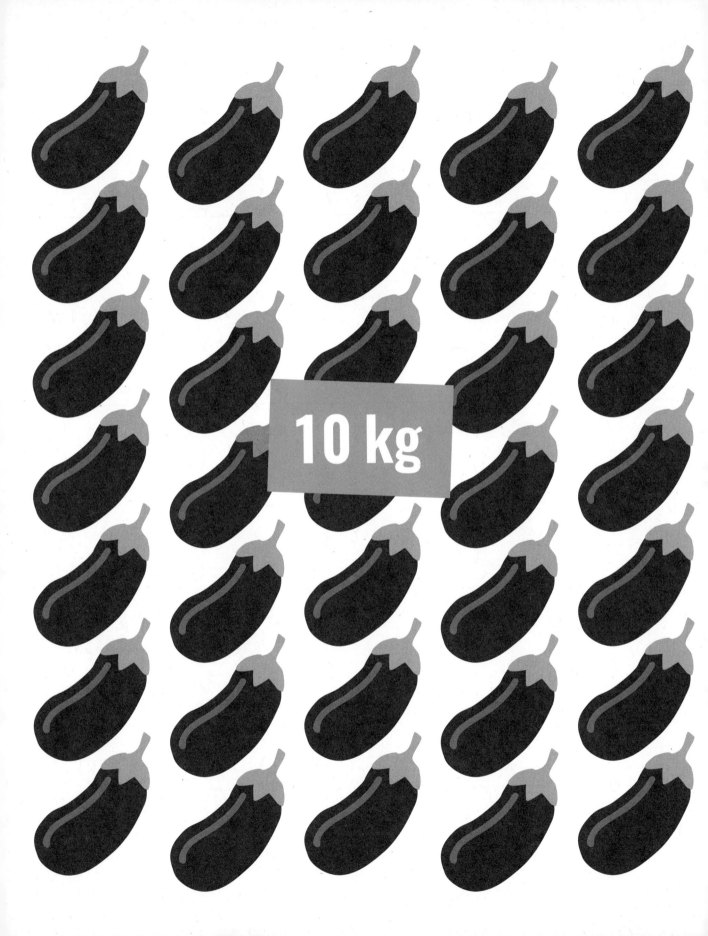

A quick and easy way to stop smoking

Can aubergines be substituted for cigarettes? Researchers have discovered that this member of the nightshade family contains nicotine, as do tomatoes and potatoes. Around 10 kilograms of aubergines contain the 1 milligram of nicotine that is consumed by smoking the average cigarette. The difference is that the substance doesn't end up in the lungs and the circulatory system but rather in the liver, where it is filtered out.

Consumption of substances harmful to the ozone layer

In the 1980s, 24 countries and the EU had a really good idea – or maybe they were just panicking – and imposed limits on the production of chemicals like fluorocarbons, which damage the ozone layer and were commonly found in aerosol cans, fire extinguishers and refrigerators.

At first glance, the initiative seems to have worked like a charm: most countries around the world are now green in this regard. The ozone layer in the earth's upper stratosphere and around the poles has recovered in many places.

However, between the 60 degrees of latitude north and south, the concentration of ozone in the atmosphere has continued to decline. No one knows why – and whether it is a result of climate change.

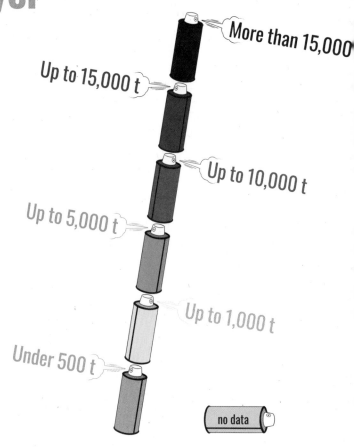

Recorded Ozone Depletion Potential (ODP), which measures the harmfulness of a substance to the ozone layer. The ODP enables us to compare different substances that cause great damage to the ozone layer.

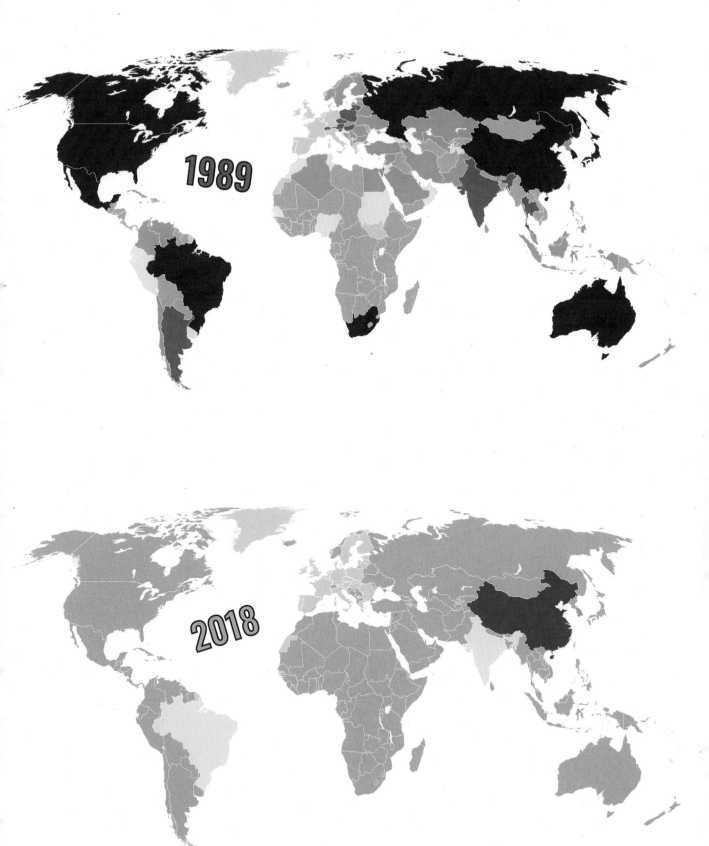

The Forests of Europe

Forests may be in decline around the world but not in Europe. The square area of woodlands has remained constant for decades and is currently even increasing by 0.5 per cent every year. That wasn't always the case, of course. Europe was most heavily deforested in the Middle Ages, when people needed land for farming, and wood was the only source of energy for a growing population.

What the world would look like if sea levels rose by 8,800 metres

Scientists never tire of calculating what would happen if sea levels were to rise by one or two metres. Goodbye, Netherlands! But what if sea levels were to rise by 8,800 metres? That can't really happen because there's not enough water on earth – but if there were, you'd have to be a mountaineer to survive.

• Everest Island

Trees as national symbols – official and unofficial

Most of our forests today are either artificially planted or affected to some degree by human activity. There remain very few primeval European woodlands. The largest one, the Białowieża, stretches between Poland and Belarus and covers 1,500 square kilometres.

Despite the fact that – or perhaps precisely because – Europeans cut down much of their forests, they often feel a close cultural connection to trees. Trees are much more than wood; they are also symbols. After the Franco-Prussian War of 1870–71, internationalists in both countries planted oaks in the hope that peace would outlive the trees. Oak trees can survive for more than 800 years. The next war between France and Germany began 43 years later.

How much space do we need to supply the entire world with solar power?

And how big is the area we need to supply the whole world with wind power?

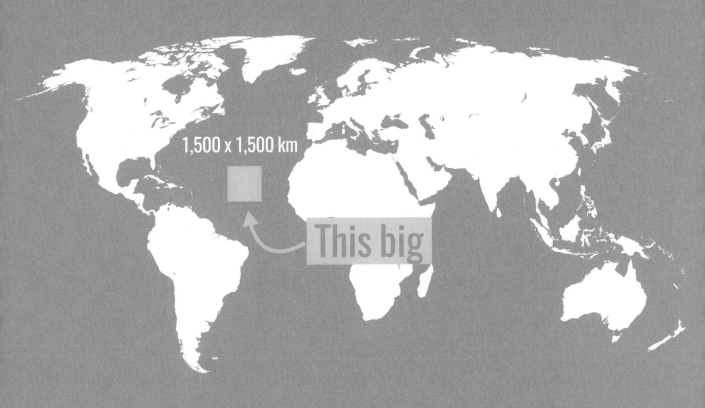

European countries that produce fewer greenhouse emissions per head than the UK

Once it became clear that greenhouse gases cause global warming, the international community agreed to reduce emissions. The biggest polluters are coal-burning power plants. In the past twenty years, Britain has concentrated on building offshore wind-power facilities. Their share of the total electricity produced in the UK is around 20 per cent.

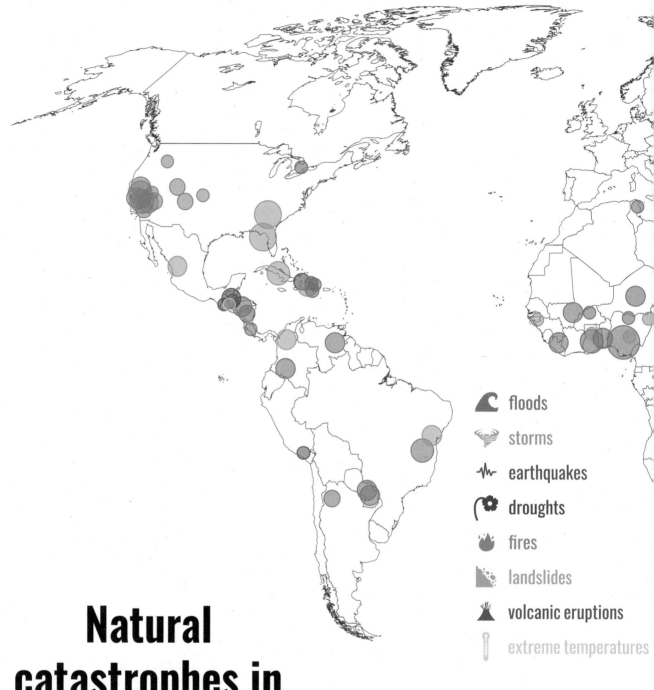

Natural catastrophes in 2018 and numbers of people affected

- floods
- storms
- earthquakes
- droughts
- fires
- landslides
- volcanic eruptions
- extreme temperatures

We have some bad news for Vanuatu. This island nation in the South Pacific is at the very top of the list of countries most at risk of being struck by natural catastrophe. According to a study of 172 countries in 2018, there are nine island nations among the top 15. The main dangers are floods, hurricanes and rising sea levels.

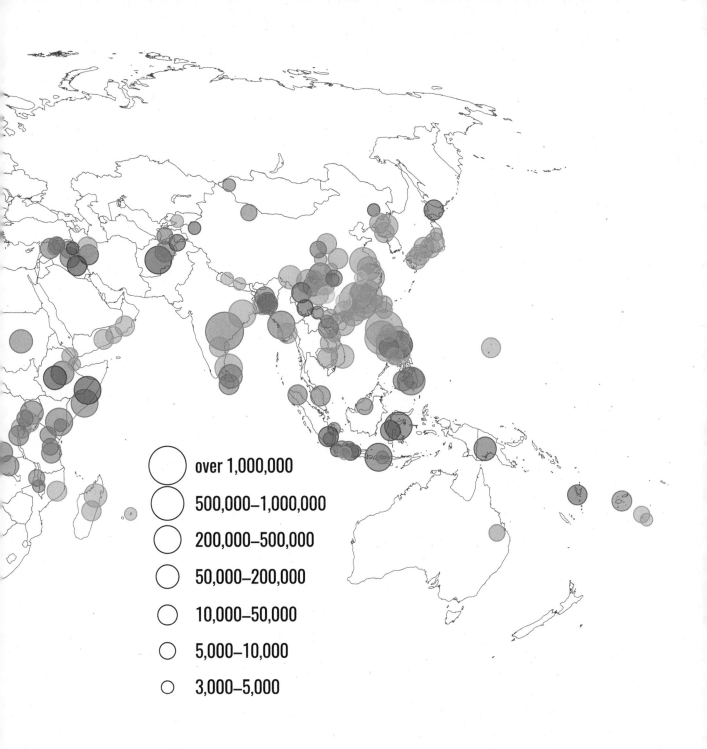

How much tropical forest does the earth lose every year?

In 2018, 12 million hectares of rain forest were destroyed – by comparison, England covers an area of 13 million hectares. The destruction of forests is particularly bad in Brazil, Bolivia, Columbia and Peru. The chief reasons for the deforestation of the Amazon region are cattle farming, palm oil production and the planting of soya beans for animal feed.

An area almost as big as England

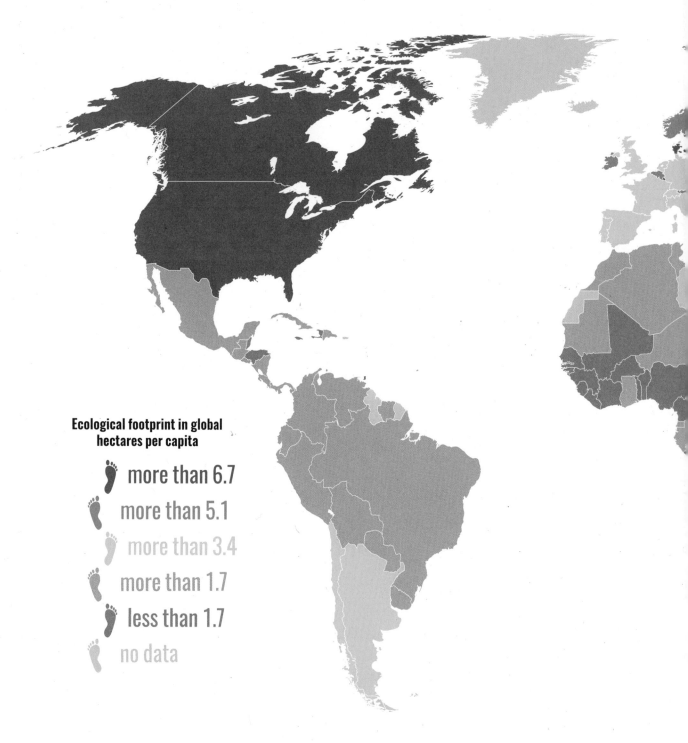

How much space per person do different countries require to maintain their typical lifestyle?

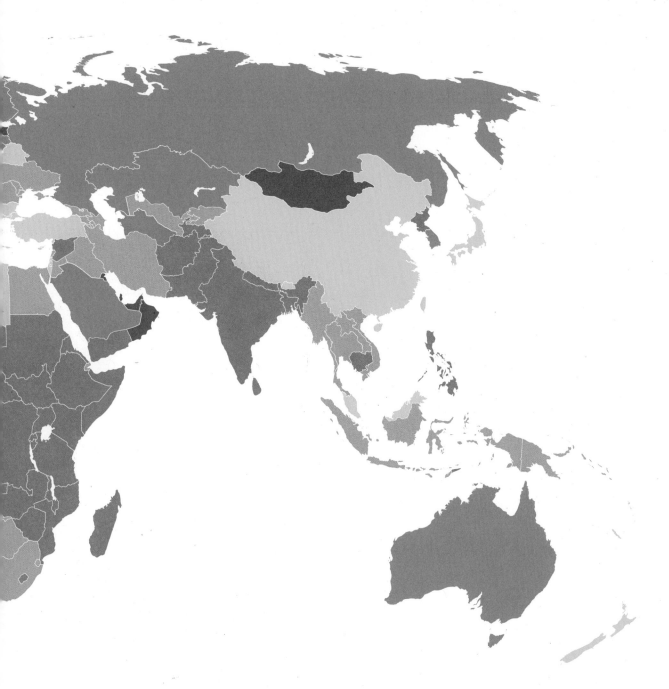

Since the 1980s, the human race has consumed more resources than the earth can provide without damage. Every person has their own ecological footprint. This is the area required to support their lifestyle and it is measured in global hectares, i.e. hectares of the earth's surface with average levels of biological productivity. (A hectare of rain forest, for example, is far more productive than a hectare of desert.) The space available on our planet is limited, of course: statistically, there are only 1.7 global hectares per person. When countries exceed this ratio, their citizens are using more resources than the earth can sustain in the long term. In 2019, the world's population used almost double the available global hectares.

Forest growth in Europe

As far back as the Romans, people in Europe ruthlessly deforested much of their land. As a result, by the start of the twentieth century, there were few woodlands left in many European countries. But that has begun to change. The square area of European forests is growing after many countries started replanting trees after the Second World War. Britain and the Netherlands have both increased the share of woodlands in their square area from 2 per cent in 1900 to more than 11 per cent at present.

 Wooded areas in 1900

 Growth by 2010

in the EU 27 and Switzerland

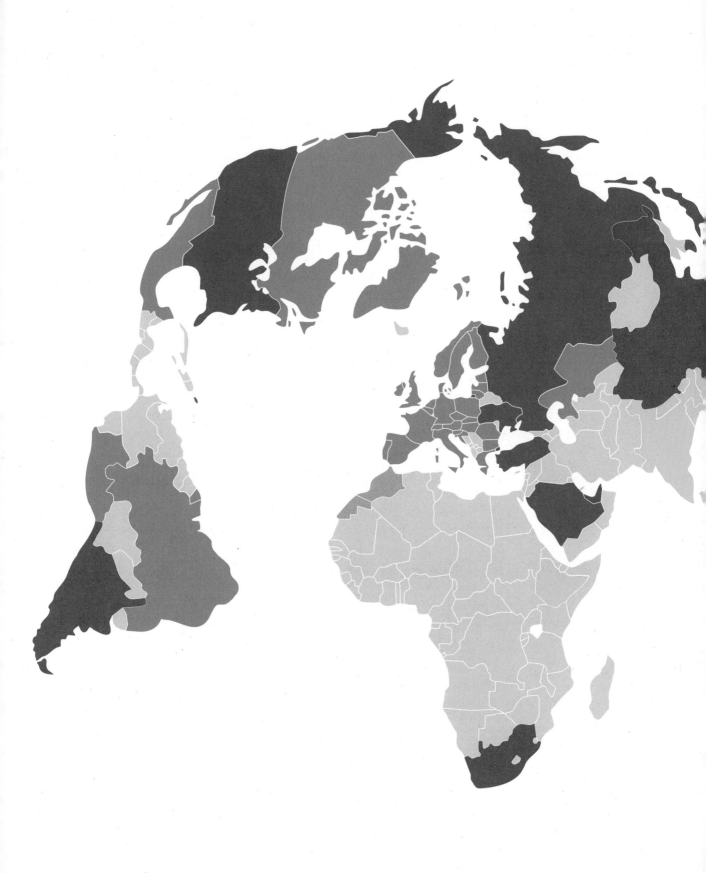

How much progress have various countries made towards fulfilling the Paris Agreement?

In 2015 politicians from 195 countries got together near Paris and decided to take joint action to save our planet – or at least ensure that it would be destroyed a bit more slowly. The idea was to reduce CO_2 emissions to curb global warming, to discuss climate problems frankly and transparently, and to help developing countries. The goal was to restrict the rise in global temperatures to 1.5 degrees Celsius. It sounded like a good plan. The thing is, not all countries are living up to their commitments.

- better than agreed in the Paris Agreement
- 1.5 degree goal achieved
- 2 degree goal achieved
- insufficient
- badly insufficient
- critically insufficient
- no data

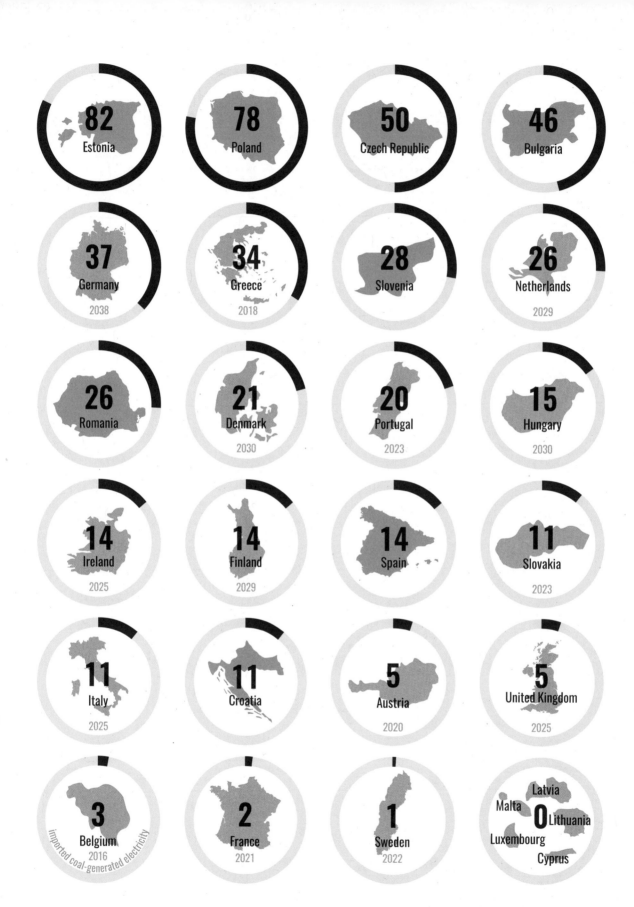

What percentage of electricity was produced in Europe by burning coal in 2018?

Estonia is often called the 'dirty devil' of Europe. Although many Estonians travel by train, the railways are powered by energy that damages the environment. The Baltic state relies heavily on oil shale, which falls into the coal category and has an even worse energy efficiency rate. Burning it leaves behind heaps of poisonous ashes. The reason for using it is obvious: almost a fifth of Europe's oil shale can be found in Estonia.

Poland's electricity also largely comes from burning coal. The country is one of the largest coal producers in the world – some 100,000 jobs depend on mining coal. France on the other hand avoids producing energy from coal; instead it relies on 56 nuclear power stations. Sweden uses more oil and gas than coal, and here the proportion of renewable energies is particularly high.

coal phase-out planned   no coal phase-out planned

For comparison

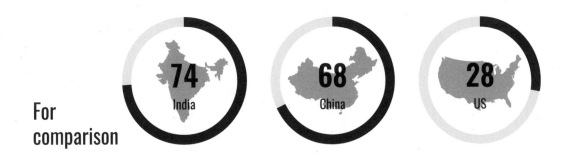

74 India — 68 China — 28 US

European cities renamed as those cities whose current temperatures they will experience by 2050

It is hard to comprehend what abstract numbers like average global temperatures actually mean in everyday life. What does an increase of two degrees feel like? The best way to explain it is by comparison. And what are people most familiar with? The cities in which they live. For that reason, a group of researchers calculated the expected changes in temperature and precipitation for 2050 in 520 big cities and altered the city names on a European map accordingly.

The cities in the northern hemisphere are getting warmer, while the tropical south is getting less rain, and the climate in general is becoming subtropical. In concrete terms, this means that temperatures and rainfall in London will become what they are today in Barcelona. Stockholm will be like today's Budapest, and Berlin will resemble San Marino. However, 22 per cent of the cities examined will have climates that cannot be compared with any presently existing cities and thus were allocated no new names.

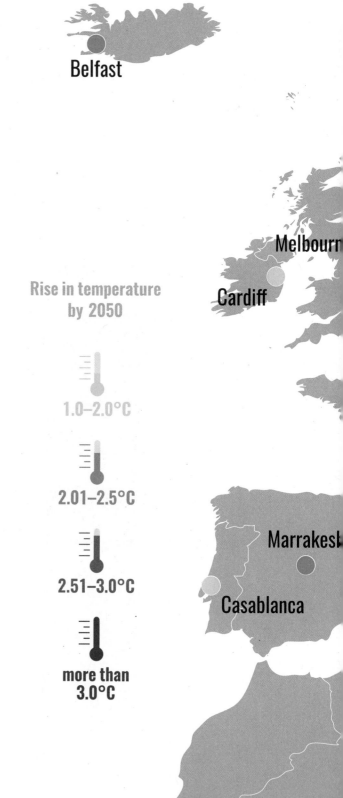

Global energy consumption between 1800 and 2017

The human race requires more and more energy, and from the twentieth century its use of primary energy sources has exploded. Such sources include coal, oil and natural gas, which are often converted into fuel rather than being used directly. It is only during major crises that energy consumption tends to fall slightly.

140,000 TWh

120,000 TWh

100,000 TWh

80,000 TWh

60,000 TWh

40,000 TWh

20,000 TWh

1800   1850   1900

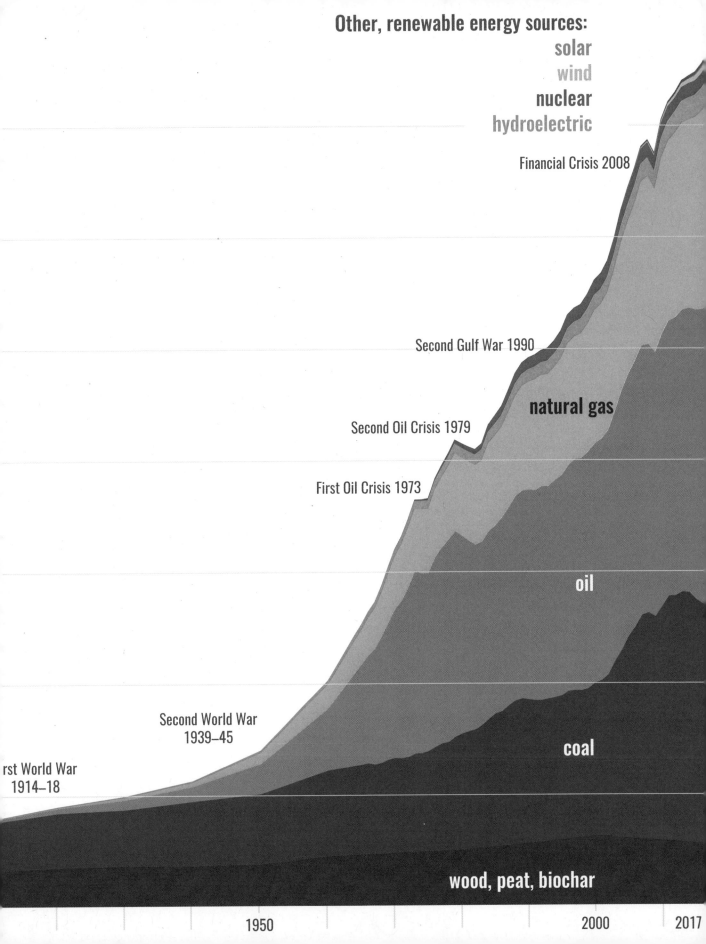

Access to Electricity in 2017

- <10%
- 10–19.9%
- 20–29.9%
- 30–39.9%
- 40–49.9%
- 50–59.9%
- 60–69.9%
- 70–79.9%
- 80–90%
- >90%

Rich in natural resources but without electricity – millions of people in sub-Saharan Africa don't have electricity. In Uganda, only 10 per cent of the population is hooked up to the power grid. Yet electricity is an indispensible condition for economic development. In North Korea too there is not enough of it available to meet the needs of industry and people.

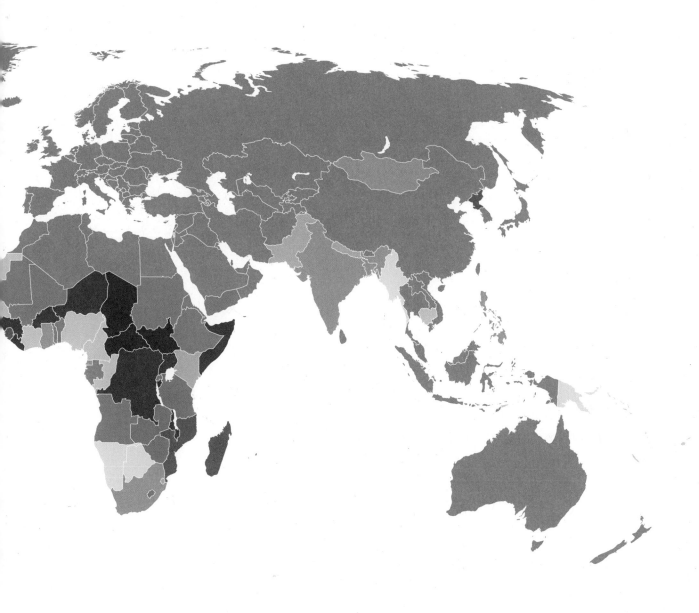

The world's largest island of rubbish is more than five times the size of the British Isles

Great Pacific Garbage Patch

Millions of tons of plastic float around in the world's oceans. In the North Pacific, this garbage has coalesced into an amorphous thing that is sometimes described as a carpet but is really more like a soup consisting of petrol cans, bottles, pieces of furniture and drinking straws. But that's only the tip of the iceberg. The plastic that floats on the water's surface is just 1 per cent of the plastic in the world's oceans. Scientists believe that the rest sinks down to lower depths.

How much have temperatures risen in European cities since 1960?

Greece is still warmer than Finland. But that doesn't mean that temperatures in these countries have stayed the same. All European cities have grown significantly warmer since 1960, even if temperatures in Greece have only risen 1.5–2 degrees compared to 4 degrees in Finland. Nothing stays the same, and it's getting hotter everywhere.

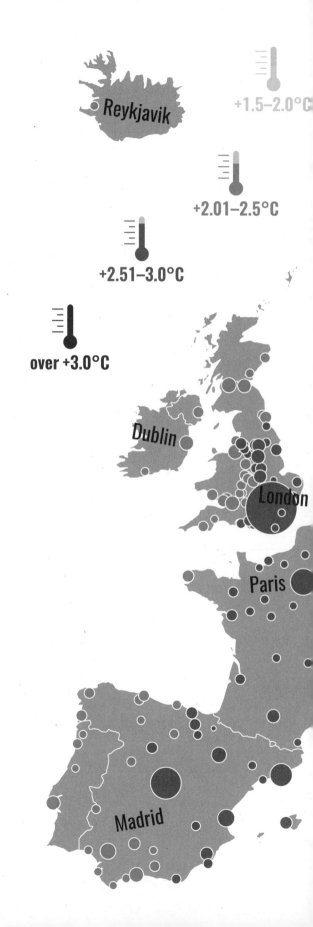

Things that burn

Things we don't like burned

Habitats of African elephants

Elephants come in two types: those with big ears and those with small ones. The big-eared variety comes from Africa, the small-eared one from Asia. Originally African elephants lived all over the world's largest continent, from the Mediterranean to the Cape of Good Hope, as we know from prehistoric cave paintings and historical reports. Today, elephant habitats form a patchwork across Africa. There are two reasons for this: changes in climate since the last Ice Age and the rise of *Homo sapiens*, who hunted elephants for their ivory and destroyed their natural habitats as the human population expanded.

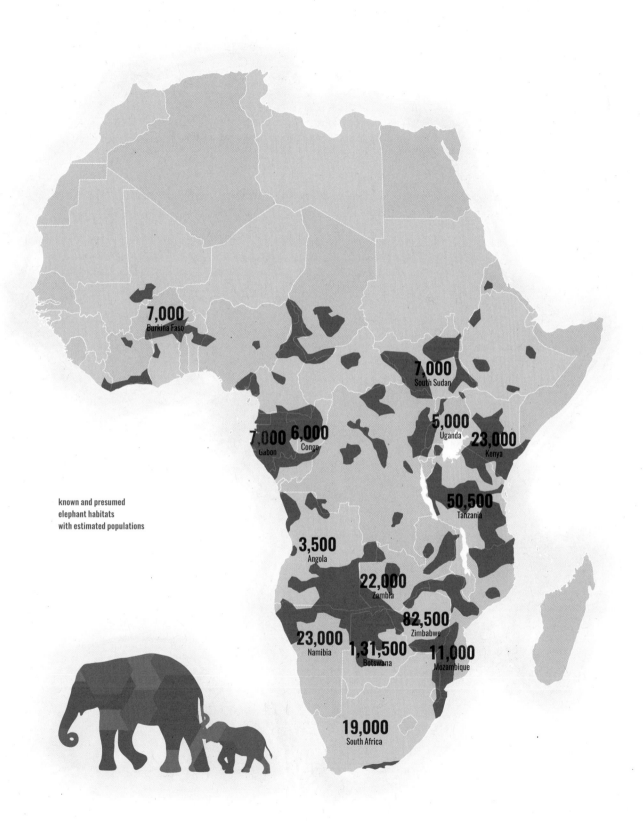

Most common source o

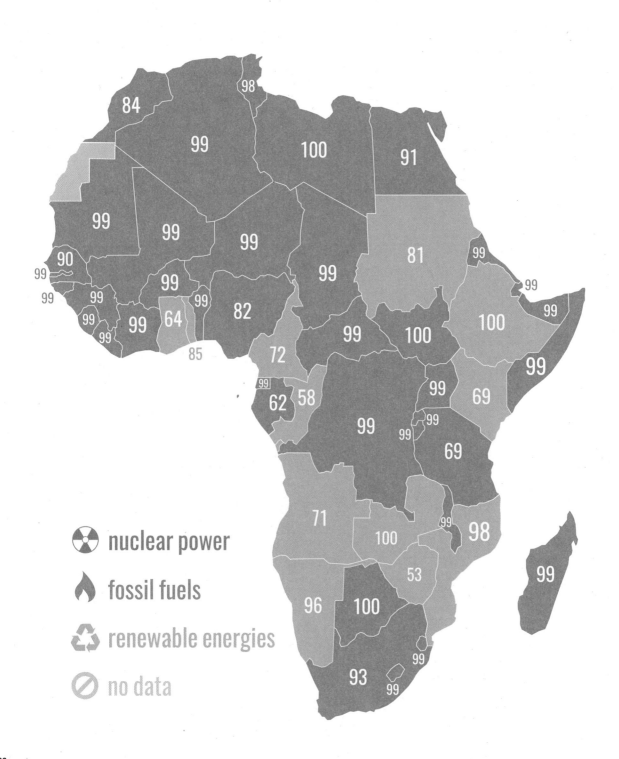

electricity in per cent (2017–18)

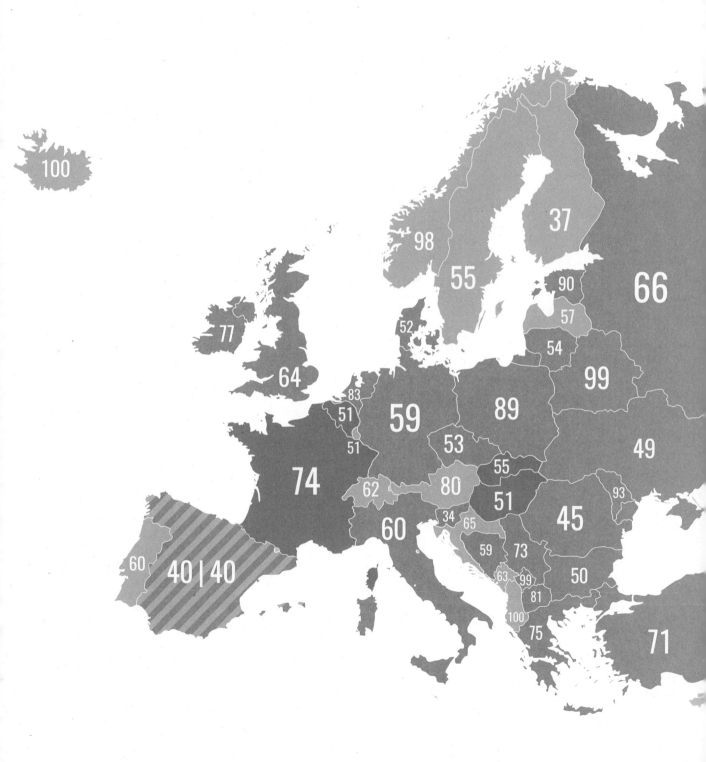

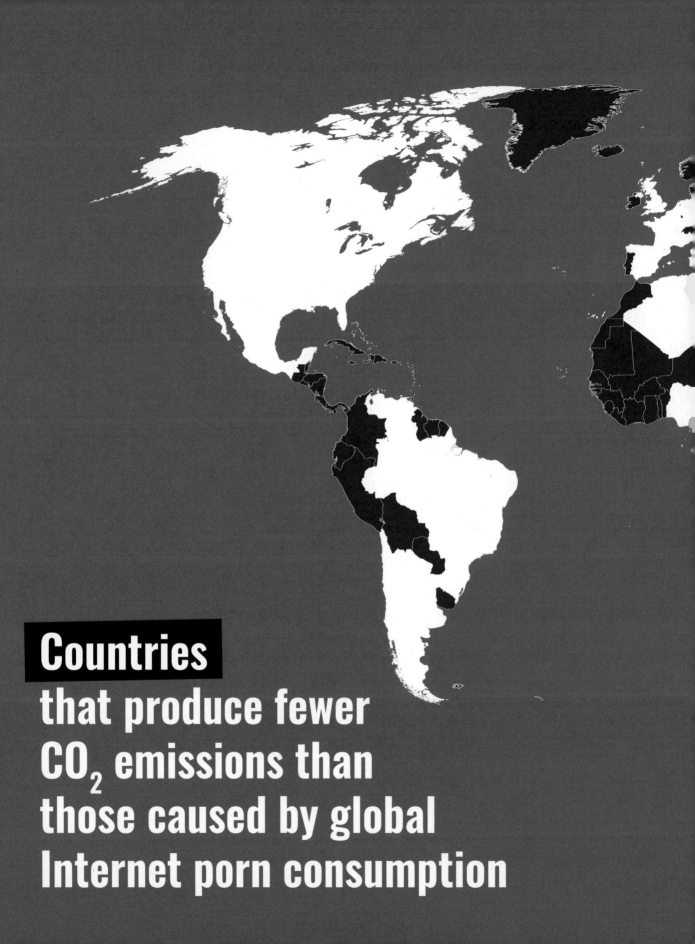

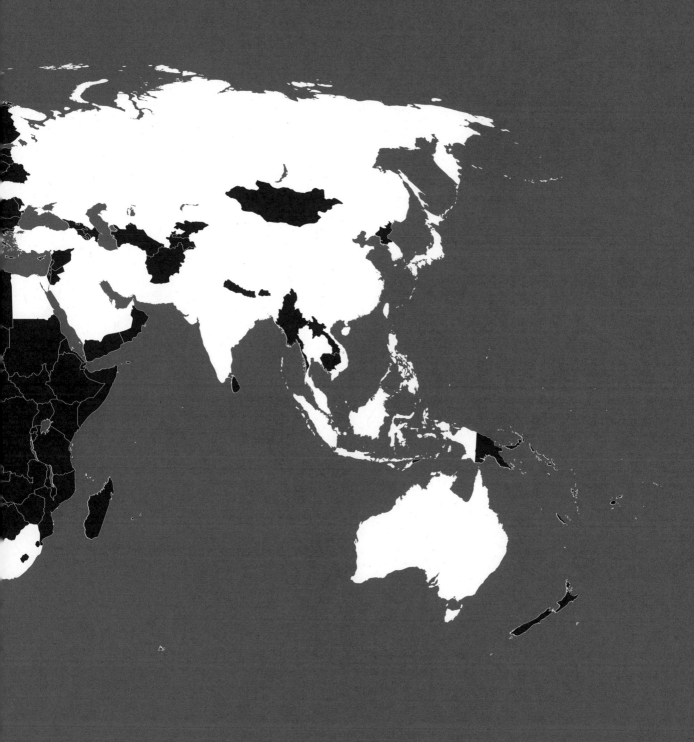

Online video streaming creates 300 million tons – or 1 per cent – of global CO_2 emissions every year. Much of this is accounted for by Netflix and Amazon (34 per cent) Youtube and other sharing platforms (21 per cent), social networks including TikTok (18 per cent), and Internet porn. Online sex caused 80 million tons of CO_2 emissions in 2018 – live webcams not included!

How many earths would we need if everyone lived like people in ...

The earth's natural resources are finite. Every year, Earth Overshoot Day – the date by which the human population has officially consumed more of those resources than the planet can replenish in any one year – makes it clear how problematic our situation has become. In 2019, Earth Overshoot Day came after only seven months. Various countries bear differing degrees of responsibility for this deficit. If, for example, people everywhere consumed as many resources per capita as people in Qatar, we would need the equivalent of 9.3 earths to meet the annual need. American and British lifestyles would require 5 and 3.7 earths respectively. By contrast, Pakistanis live well within their planetary means, requiring only 0.4 earths.

Qatar

United States

United Kingdom

Pakistan

Tree rings

Fingerprints

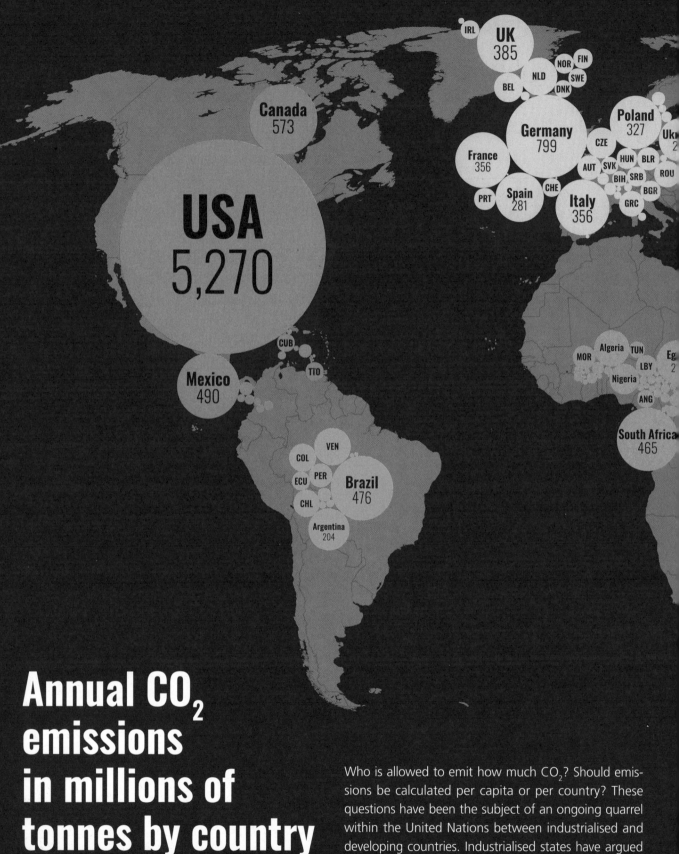

Annual CO$_2$ emissions in millions of tonnes by country

Who is allowed to emit how much CO$_2$? Should emissions be calculated per capita or per country? These questions have been the subject of an ongoing quarrel within the United Nations between industrialised and developing countries. Industrialised states have argued

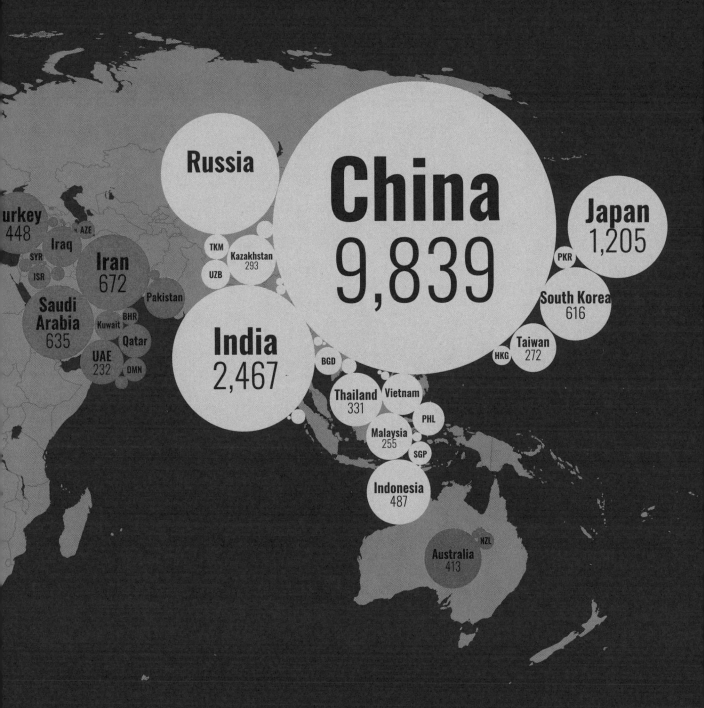

for decades that one of the world's major developing nations, China, with its large population and rapidly growing economy, is producing too much CO_2. And indeed, China is responsible for the greatest amount of emissions of any one country in the world. Detractors therefore agree that China is damaging the climate more than anyone else and must be prepared to reduce its emissions.

Developing countries couldn't disagree more. Not only do industrialised countries produce more CO_2 per capita than China; they have also damaged the environment far more in the past. Seen in this light, developed nations' demands on China seem hypocritical.

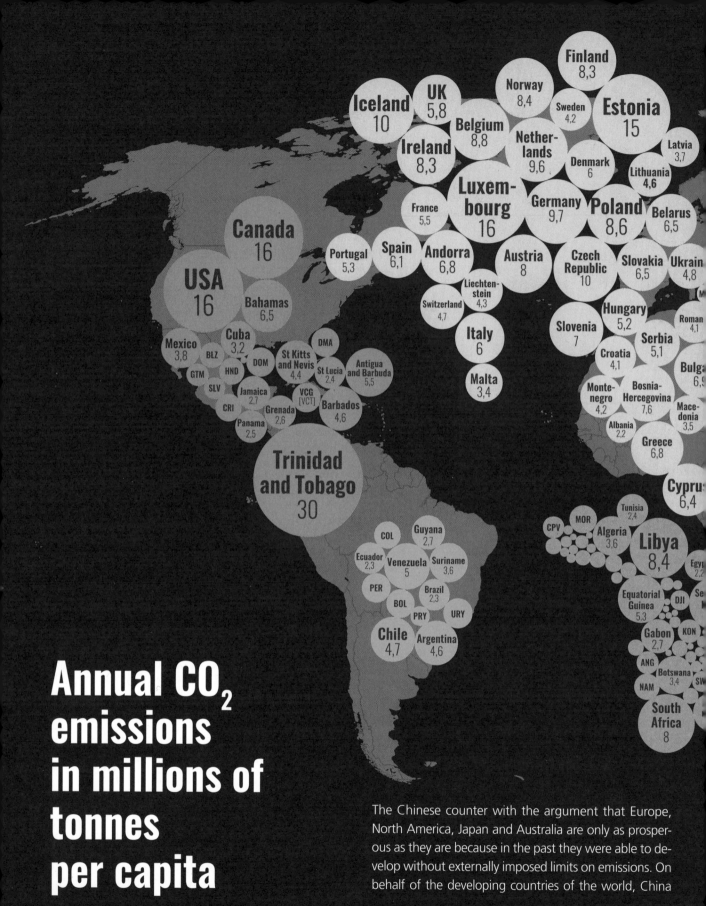

demands the same freedom and insists on deciding for itself which energy sources are best for its economy. This is an understandable position. Coal-fired power plants are a major economic growth factor in many countries. At present, countries like Germany and Austria still emit more CO_2 per capita than China – to say nothing of the US. As long as this is the case, Western calls for Chinese environmental action are likely to remain unheeded. Things are different in Switzerland. There, for the sake of climate neutrality, CO_2 emissions are capped at one tonne per capita per year.

Number of actively used mobile phones worldwide

8 Billion

Number of actively used toothbrushes worldwide

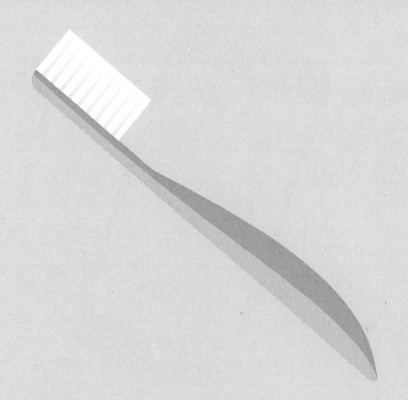

3.5 Billion

Usage of plastic cups in Europe in 1789

Everything was better in the past – or almost everything. The human race had to wait until 1907 before the disposable paper cup was invented in the US. And the wait was even longer before we could drink our coffee from a plastic cup.

Disposable cups weren't originally intended for drinking coffee. They were designed for water coolers in the US, since at the beginning of the twentieth century water was being promoted as a healthy alternative to alcohol. When a little later the Spanish Flu claimed millions of lives the world over, hygienic advantages became a powerful argument for producing disposable containers.

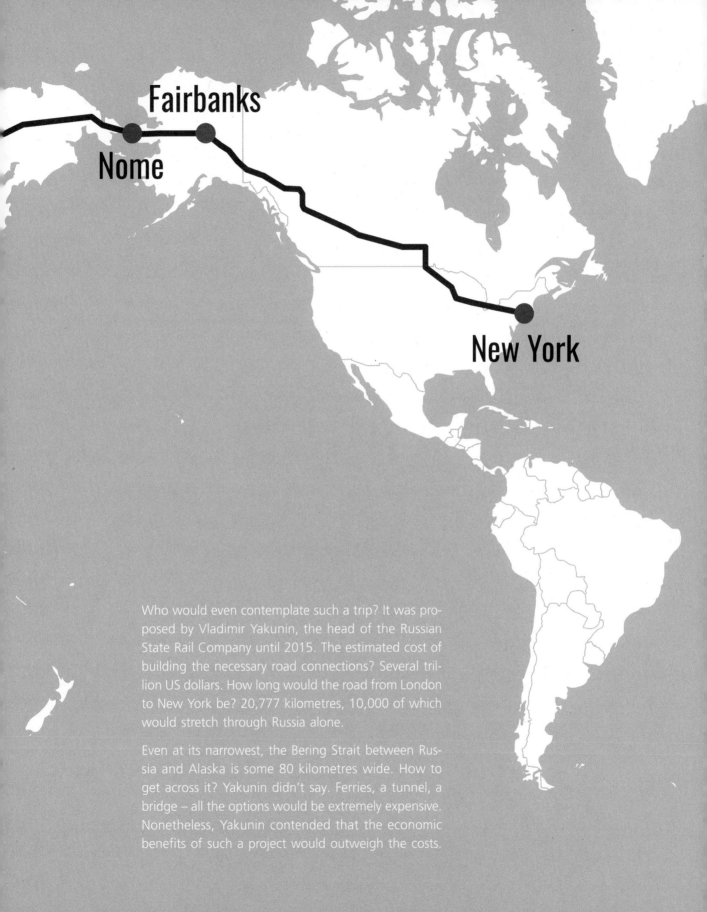

Who would even contemplate such a trip? It was proposed by Vladimir Yakunin, the head of the Russian State Rail Company until 2015. The estimated cost of building the necessary road connections? Several trillion US dollars. How long would the road from London to New York be? 20,777 kilometres, 10,000 of which would stretch through Russia alone.

Even at its narrowest, the Bering Strait between Russia and Alaska is some 80 kilometres wide. How to get across it? Yakunin didn't say. Ferries, a tunnel, a bridge – all the options would be extremely expensive. Nonetheless, Yakunin contended that the economic benefits of such a project would outweigh the costs.

How much of the earth's surface has been paved over?

Human beings have paved over 1.18 million square kilometres with asphalt, concrete or other manmade surfaces. That is 0.8 per cent of our planet's land surface. 27 per cent of these paved areas can be found in Europe, although the continent accounts for only 7 per cent of the world's land mass.

This much →

CO_2 emissions per litre

Bottled water

202.74 g

Does it make a difference whether you drink bottled water or tap water? If you consider CO_2 emissions, there's a massive difference. For bottled water 52 per cent of them are caused by the production and recycling of the bottles – none of which features in tap water. Further emissions are created by the distribution, transport after purchase and refrigeration of bottled water. For comparison, driving your car for one kilometre creates 200 g of CO_2.

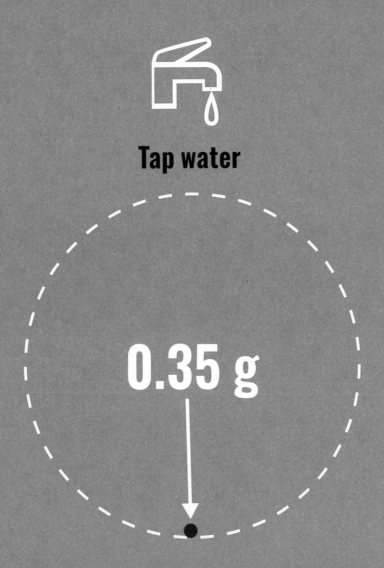

Tap water

0.35 g

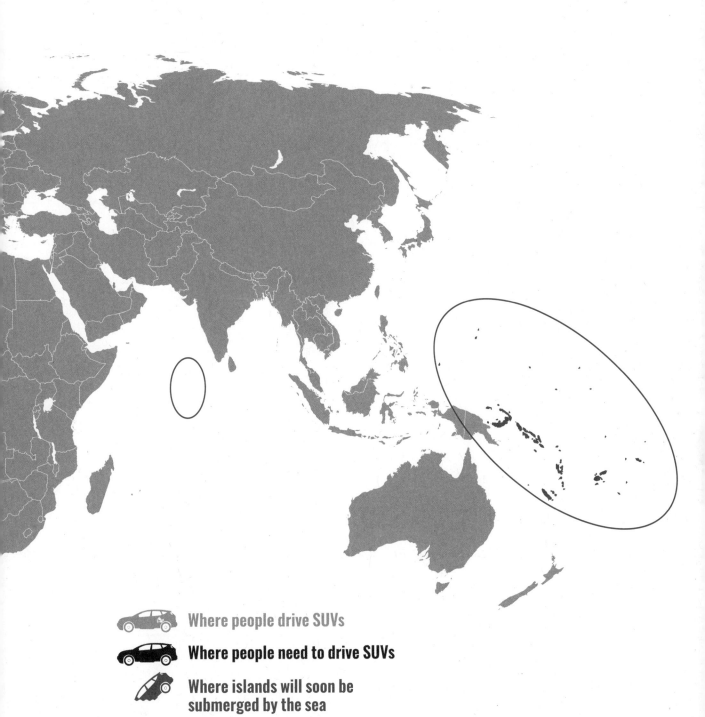

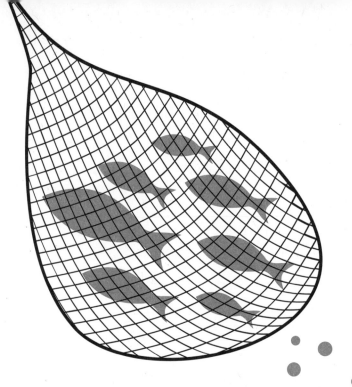

Fish removed

and rubbish added

(worldwide per minute)

Every year, as much as 10 million tonnes of our rubbish end up in the world's oceans – that's about 19 tonnes per minute. One major polluter is the fishing industry. Around 10 per cent of plastic waste in the oceans consists of abandoned fishing nets, buoys, lines, traps and cages. Conversely, in 2017, almost 92.5 million tonnes of fish were extracted from the planet's oceans – approximately 176 tonnes every minute. And that doesn't include aquaculture and freshwater fishing.

Desertec plans for the distribution of renewable energy

One day, when representatives from the biggest energy and financial companies in Europe and Africa got together to try to save the world, they imagined a 300-square-kilometre area equipped with solar power facilities capable of supplying the entire globe with electricity. The sun would be the energy source; and in the desert, damage to natural habitats would be almost zero. In addition, their plans foresaw hydroelectric power generated in the mountains, and wind power in Europe's breezy north.

The plan was called Desertec but it disappeared without a trace in 2014, when the Arab Spring intervened. But that didn't stop nations in northern Africa from realising some of the ideas. The world's largest solar power installation is currently under construction in Morocco, for example. It will have a capacity of 2,000 megawatts, enough to supply 700,000 private households in Europe with electricity.

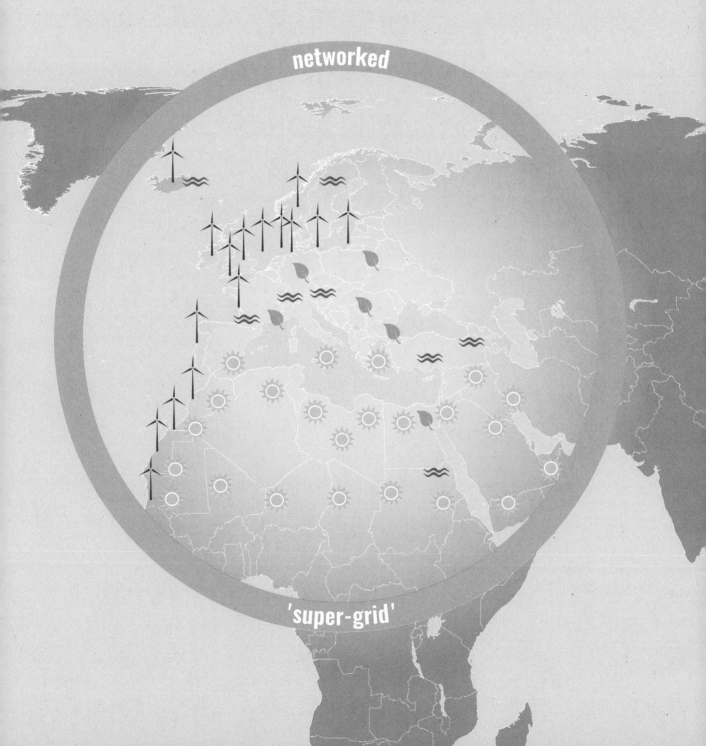

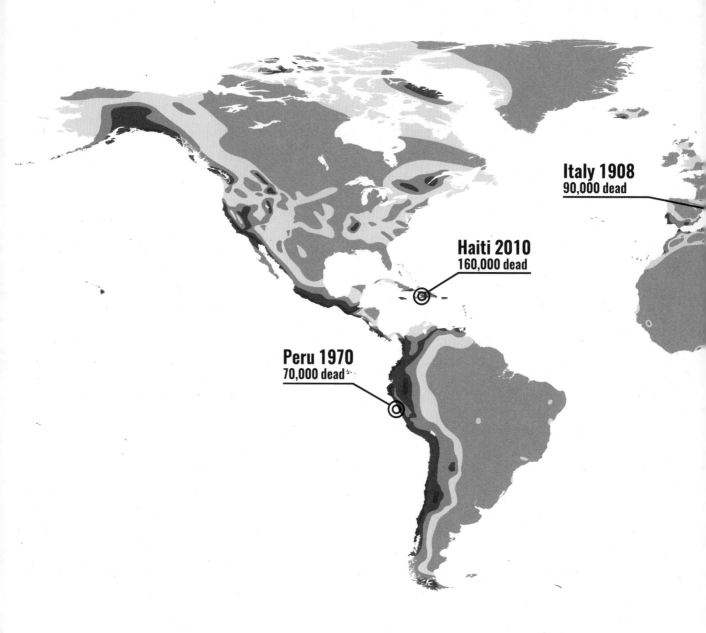

Earthquake risk zones and selected earthquakes with high casualties since 1900

Casualty figures are estimates and differ according to sources used

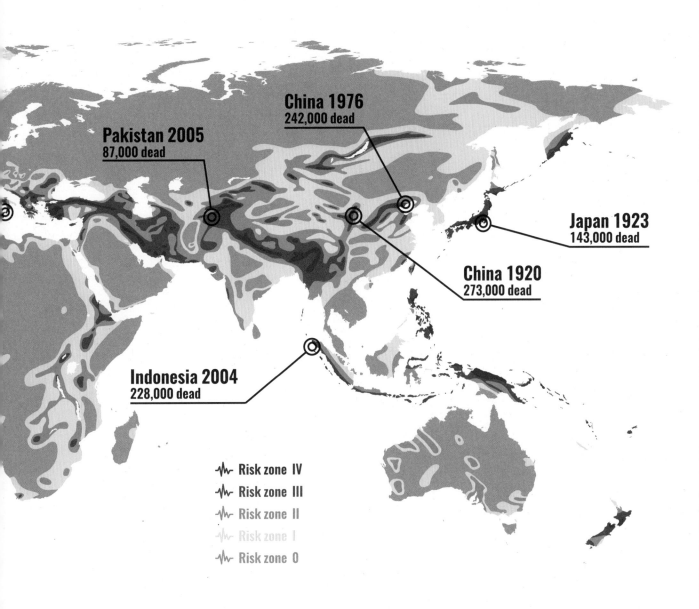

Three times a week, there's an earthquake somewhere on our planet, with the most affected countries being Indonesia, the Philippines, Japan, the US, Mexico, Peru and Chile. One researcher has concluded that people in places with a prevalence of catastrophes – as long as they are unpredictable – are particularly religious. This innate tendency towards religiosity can be successfully exploited by missionaries, as the prominence of Christian relief organisations in risk areas suggests.

Portland's path to becoming a bicycle city

Cycling accounts for up to 7 per cent of traffic in Portland, Oregon, with its 650,000 inhabitants. Bikes are truly booming in the city. Over the last 20 years, the number of cyclists has increased by 700 per cent. The reason is that the city has invested in cycle roads, bike traffic lights and cycle-friendly intersections. In the past 40 years, 600 kilometres of new cycle paths have been built.

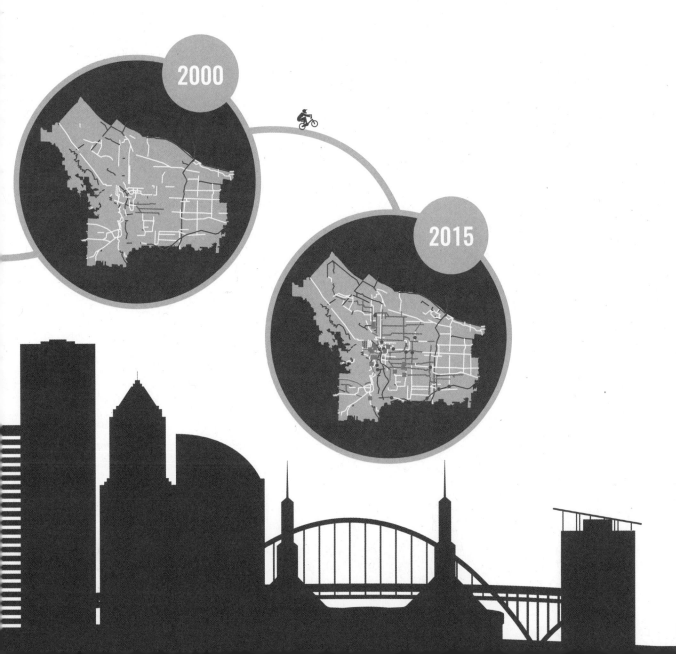

Percentages of male and female smokers

In Russia, around a third of the population smoke, despite a legal age of eighteen and warnings about the health risks on all tobacco products. All in all, 59 per cent of Russian men and some 23 per cent of women smoke. By the way, smoking is officially prohibited on the Trans-Siberian Railway but is tolerated in the passageways between cars.

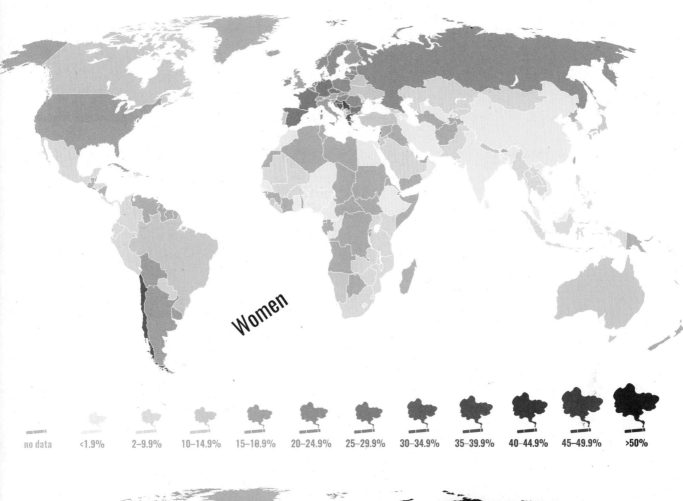

no data <1.9% 2–9.9% 10–14.9% 15–19.9% 20–24.9% 25–29.9% 30–34.9% 35–39.9% 40–44.9% 45–49.9% >50%

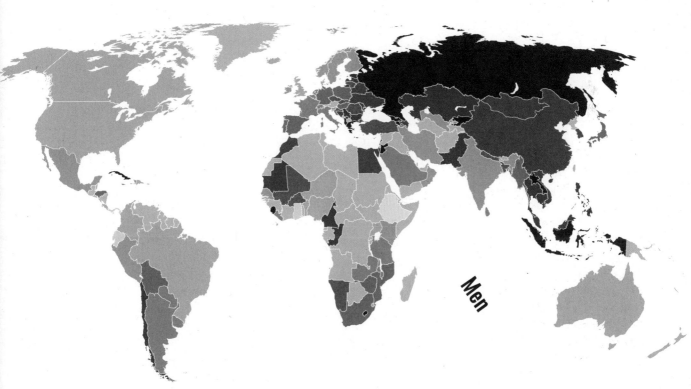

Human beings account for just 0.01 per cent of the earth's total biomass ...

... and still they behave like utter bastards towards all the rest of living things. According to the so-called Biomass Balance report of 2018, *Homo sapiens* has displaced 83 per cent of all wild mammals, 80 per cent of all marine mammals, 50 per cent of all plants and 15 per cent of all fish. One researcher described the situation like this. His daughter enjoys doing puzzles showing rhinos next to giraffes next to elephants. That's nonsense of course: a more realistic puzzle would feature a cow next to a cow next to a cow next to a chicken. Livestock animals now make up 60 per cent of all mammals on earth.

all human beings
0.06 gigatonnes of carbon

livestock
0.1 gigatonnes of carbon

Biomass of land mammals

wild animals
0.007 gigatonnes of carbon

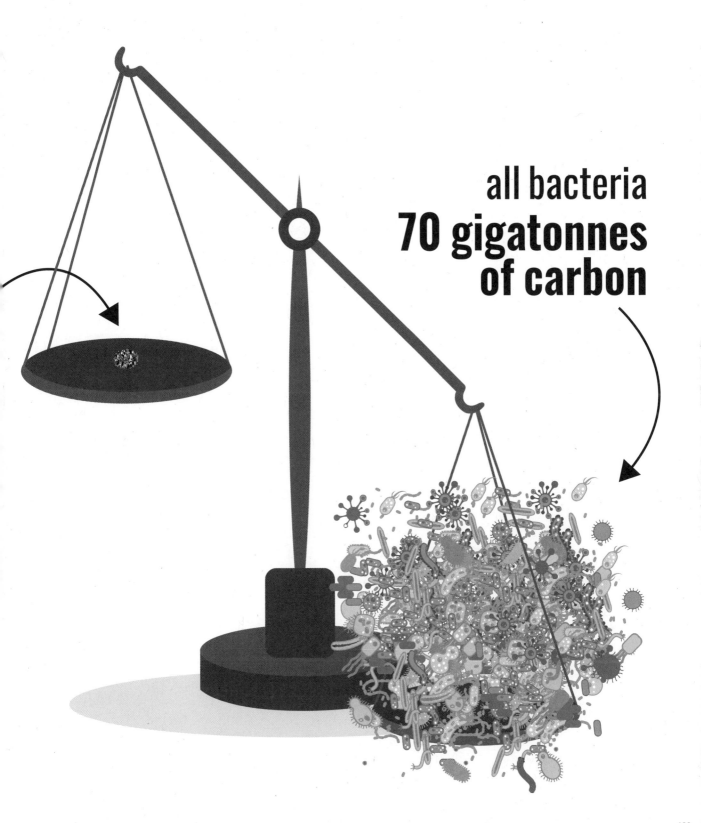

Philippine Sea
5,695,060 km²

Arabian Sea
4,232,223 km²

Coral Sea
4,035,711 km²

Bering Sea
2,433,324 km²

Caribbean
2,871,349 km²

South China Sea
3,329,135 km²

Bay of Bengal
2,195,986 km²

Sea of Okhotsk
1,658,834 km²

Mediterranean
1,646,162 km²

Tasman Sea
3,318,494 km²

The ten largest seas by surface area

The heavyweights among bodies of water are the world's oceans: the Pacific, Atlantic, Indian, Arctic and Antarctic – although sometimes only the first three are considered proper oceans. We distinguish between them and seas, which are separated from the oceans by island chains (marginal seas) or narrow straits (inland seas).

Annual toilet paper consumption per bottom in kilos

Almost everyone uses it – but not everyone uses the same amount. Still, consumption is not just about quantity but also about quality. Using toilet paper made from recycled material can save an estimated 50 per cent of energy and 33 per cent of water over conventional bog roll.

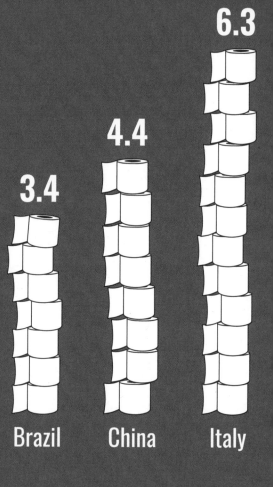

3.4 Brazil
4.4 China
6.3 Italy

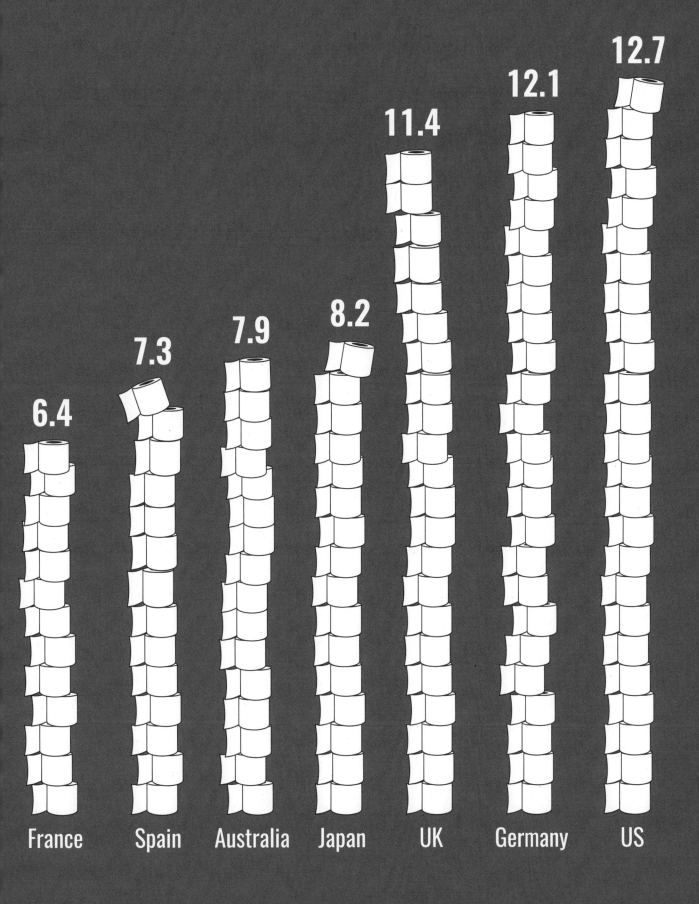

The ten worst maritime oil catastrophes

In 2010 in the Gulf of Mexico, the Deepwater Horizon drilling platform exploded. Specialists arrived from Texas, 600 kilometres away, to put out the fire, which went on for 36 hours. The resulting oil slick was the size of Jamaica and contaminated 1,000 kilometres of coastline and the bottom of the sea. Rescuers suffered from health after-effects; countless birds, dolphins and turtles died; and many people lost their jobs. No one knows how much crude oil was spilled into the Gulf, but one US court estimated it at 3.19 million barrels – or more than 500 million litres.

Deepwater Horizon (2
492,000–627,000 t

Ixtoc I (1979)
454,000–480,000 t

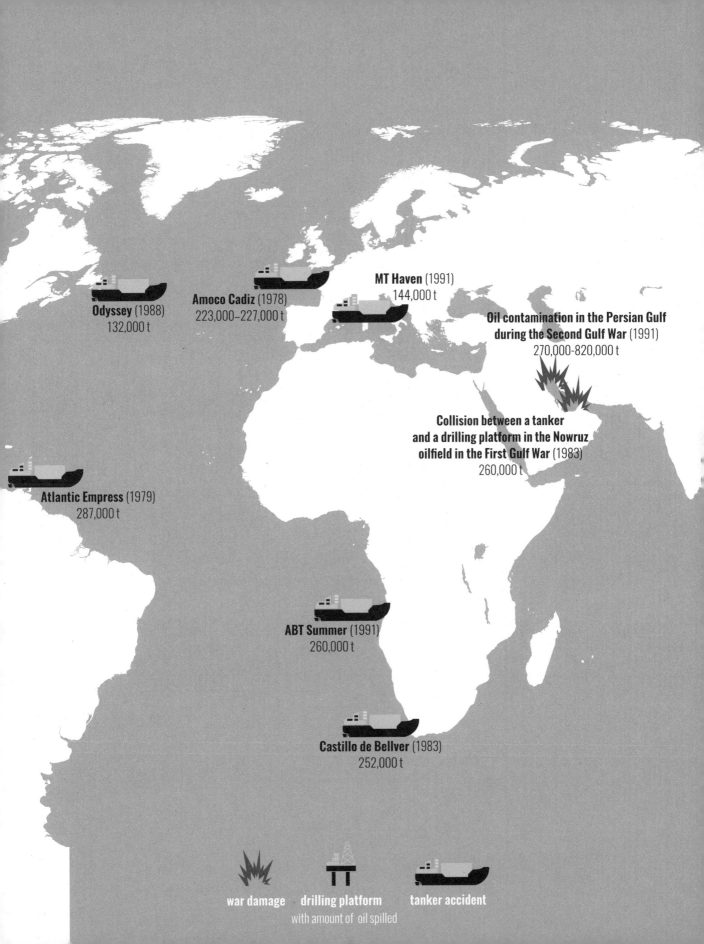

All the world's nuclear power plants

- active
- decommissioned
- under construction

**More people
live in the green
regions than in
the blue ones**

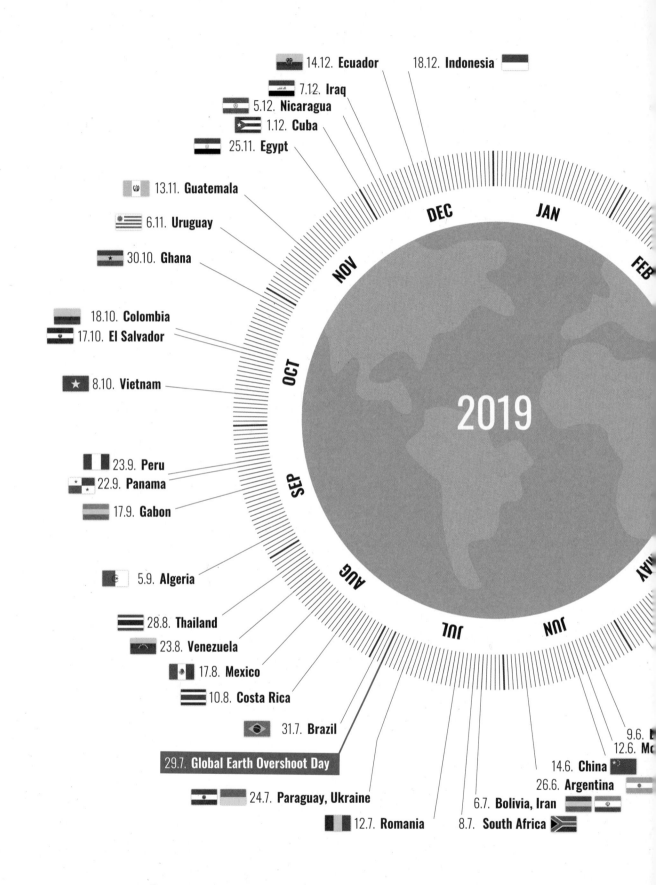

Earth Overshoot Day comes earlier every year

Earth Overshoot Day is the symbolic date after which human beings run a sustainability deficit, for example by consuming more combustible fuels or food than can be replenished or produced. In 1979, Earth Overshoot Day fell on 29 October. In 1989 it came on 11 October, in 1999 on 29 September, in 2009 on 18 August, and in 2019 on 29 July.

Individual countries have their own national Earth Overshoot Days. Poorer nations with low levels of consumption like Indonesia, Ecuador or Nicaragua only go into the red in December. In Qatar or Luxembourg, Earth Overshoot Day already comes in February. That's because both of these countries import most of their food, which causes higher CO_2 emissions and spoils the party.

- 11.2. Qatar
- 16.2. Luxembourg
- 8.3. United Arab Emirates
- 11.3. Kuwait
- 15.3. US
- 18.3. Canada
- 29.3. Denmark
- 31.3. Australia
- 3.4. Sweden
- 6.4. Belgium, Finland, Saudi Arabia
- 10.4. South Korea
- 12.4. Singapore
- 18.4. Norway
- 26.4. Russia
- 27.4. Ireland, Slovenia
- 3.5. Germany, Israel
- 4.5. Netherlands
- 6.5. New Zealand
- 9.5. Switzerland
- 13.5. Japan
- 14.5. France
- 15.5. Italy
- 17.5. United Kingdom
- 19.5. Chile
- 20.5. Greece
- 6.5. Portugal
- 5. Spain

Fires visible from satellites in the summer of 2019

In the past, satellite images like this were used to illustrate the increasing prevalence of forest fires. That's a little misleading, though, since satellite images are now so precise that they also capture small-scale fires, and what looks like a wildfire may just be an intentional, controlled industrial or agricultural fire. Nonetheless, it remains true that in 2019 a particularly large amount of woodlands burned down. This map is a rough indication of such fires in the months of July and August.

Countries renamed as other nations with similar CO$_2$ emissions per capita

The average German produces the same amount of CO$_2$ as two Ukrainians: more than 9 tonnes per year. For a private household these emissions are caused by heating (36 per cent), transport (26 per cent), food (12 per cent) and consumption of other products and services (25 per cent).

 over 10
 over 9
 over 8
 over 7
 over 6
 over 5
 over 4
 under 4

annual CO$_2$ emissions in tonnes, per capita

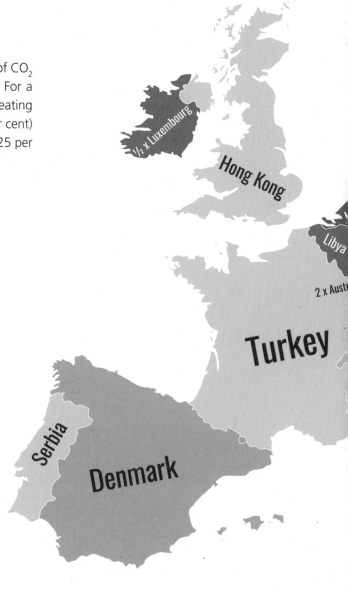

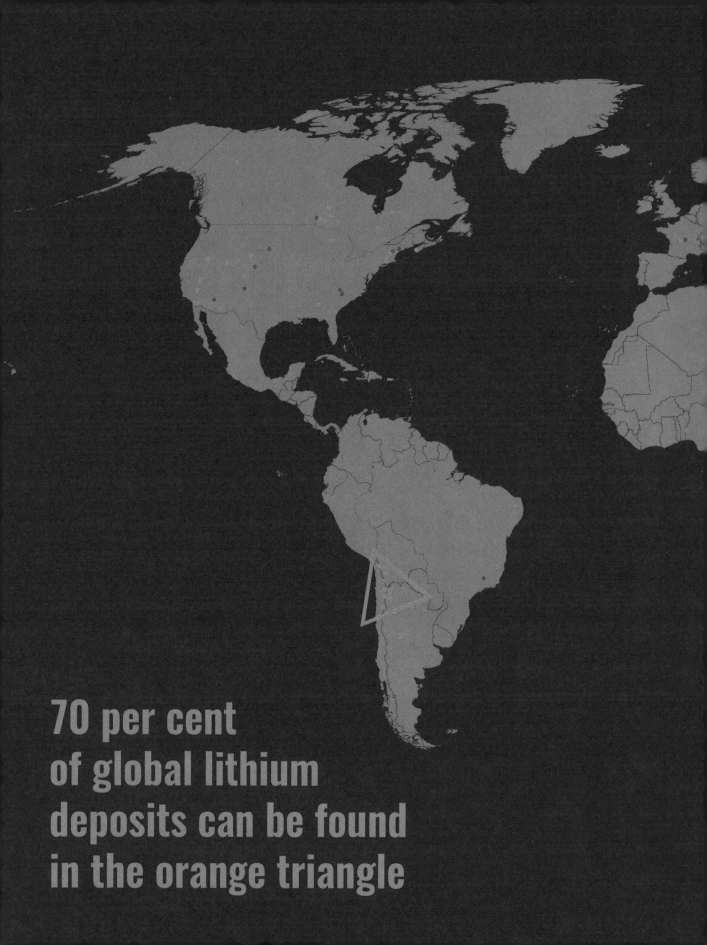

■ Lithium alone
■ Mixed deposits (lithium and other raw materials)

The transition to sustainable energy supplies requires electric vehicles. They run on batteries that need lithium. This light metal is found above all in the 'lithium triangle' between Chile, Argentina and Bolivia. The problem is that Chile's indigenous population opposes lithium mining because it would destroy their way of life, for example around the Salar de Atacama salt flats. They also object to private companies being the sole profiteers from mining. In Chile, a large number of key industries are privatised. It is the only country in the world in which the water supply is completely in private hands.

■ Countries which Germany criticises for failing to dispose of plastic waste properly and thereby contaminating the world's oceans

■ Countries to which Germany sends a large part of its plastic waste

Germany may be the motherland of waste separation, a place where people take enormous care to ensure that everything from pesto jars to egg cartons goes into the right recycling bins. But it also produces more packaging waste than any other EU country: 220 kilos per capita in 2016. In Bulgaria, the comparable figure was 57.4 kilos.

Much of Germany's waste gets recycled, some of it is burned and a million tonnes of plastic are exported. For years, most of Germany's plastic waste – along with half of all the world's – went to China. But in 2018, the Chinese government stopped the imports and since then, Germany has used other Asian countries as waste receptacles. In 2018, 7 per cent of German plastic waste was still being sent to Hong Kong, from where it was shipped on to other countries.

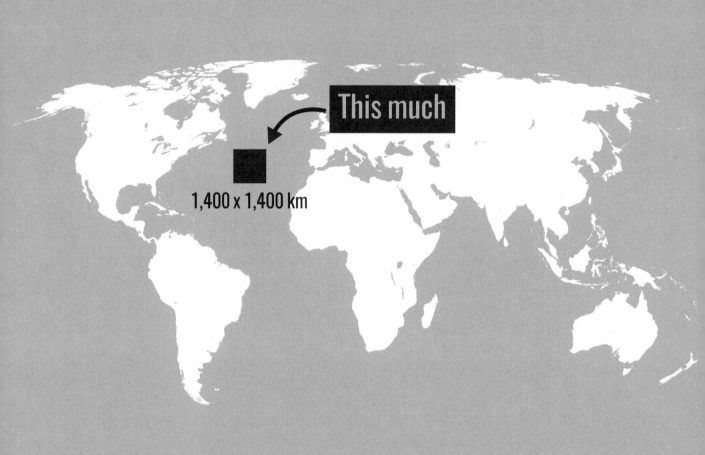

How much space would be needed to meet the global fuel demand with biofuel made of micro-algae?

Burning oil products is bad for the environment but biofuel made from sugar beets or rape seed requires lots of cultivated land. So why not produce a substitute fuel made of micro-algae? Algae don't need fresh water or arable land and thus save on CO_2. Unfortunately, algae-based biofuel is a long way from being ready for the market. What the map doesn't show is that micro-algae are usually cultivated on land. And not all algae are welcome …

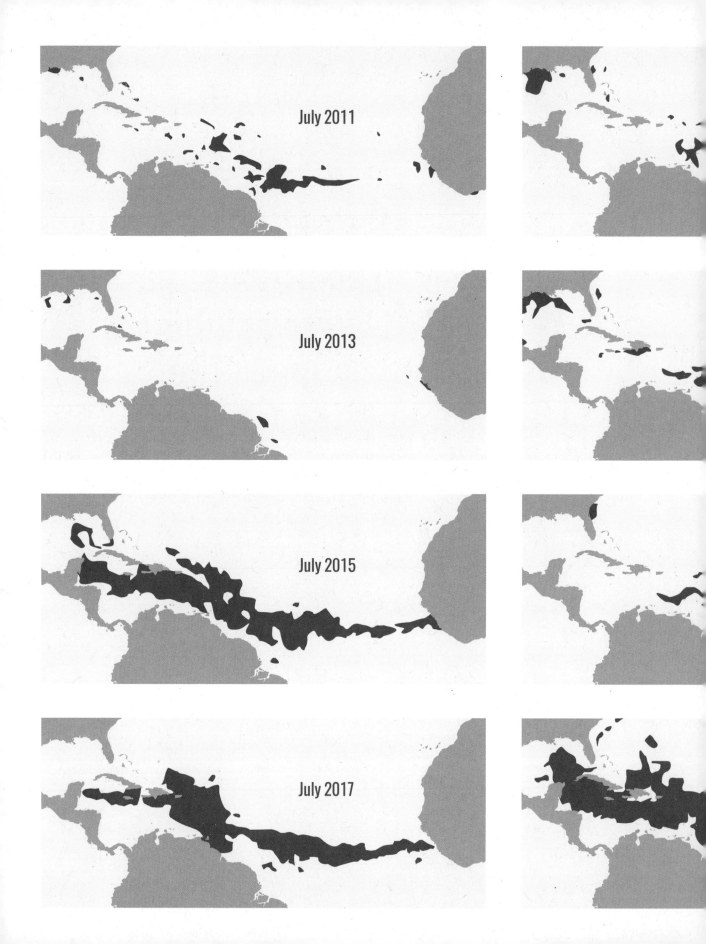

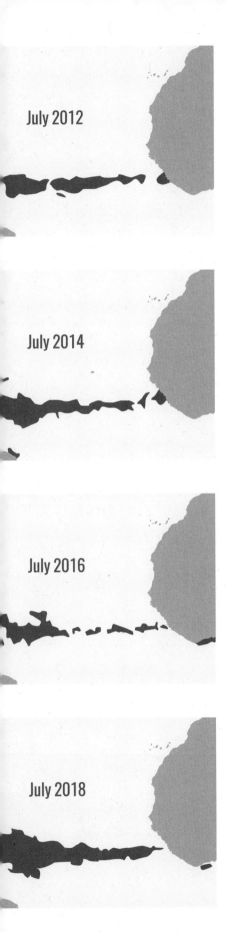

July 2012

July 2014

July 2016

July 2018

Carpet of algae in the Atlantic gets bigger and bigger

The largest carpet of algae in the world stretches from West Africa all the way to the Gulf of Mexico – 8,850 kilometres – as an American research team discovered by analysing satellite images. What the researchers don't know is how this 20-million-tonne mass of algae came about. One theory is that in winter nutrient-rich water from deep down rises to the surface before the African coast. That would encourage the growth of algae.

At the same time, after cutting down parts of the Amazon rain forest, people apply nutrition-rich fertiliser to the land that eventually makes its way into the ocean. High concentrations of algae are harmful to all other forms of marine life, which get less sunlight. That disrupts the ecosystem.

Particulate matter, manmade

With every breath we take, we inhale tiny particles that are usually manmade and come from fertilisers and burning processes in ovens, heating units or power plants. That may sound harmless but particulate

1   **50**   **100**

in µg/cubic metre of air
(PM 2.5: particulate matter with a diameter of 2.5 micrometres)

matter pollutes the air we breathe. The problem is particularly severe in China and India. There are also natural causes of particulate matter, such as forest and brush fires, volcanic eruptions and soil erosion. Concentrations of dust are particularly great in the Sahara. This map, however, only shows manmade particulate matter.

An island in a lake on an island in a lake on an island

Do you want to get away from people? It's unlikely that anyone has ever set foot on this nameless and remote island: it is the largest island in a lake on an island in a lake on an island. Before its discovery, the only third-degree island ever found on earth was in the Philippines.

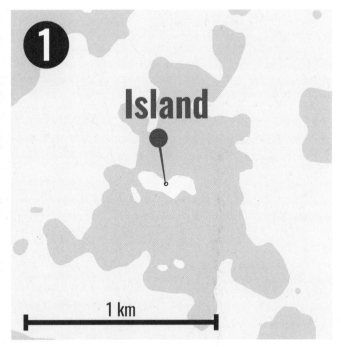
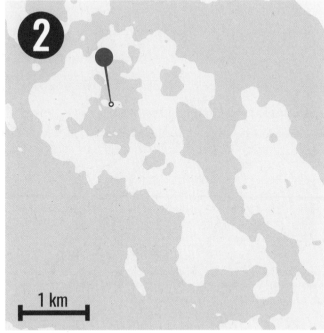

These 67 companies emitted 67 per cent of all industrial greenhouse gases between 1988 and 2015

Suncor Energy
Encana
Canadian Natural Resources
Devon Energy
Arch Coal
Consol Energy
Chevron
Peabody Energy
Anadarko Petroleum
Alpha Natural Resources
Vistra Energy
Natural Resource Partners
ExxonMobil
ConocoPhillips
Marathon Oil
Occidental Petroleum
Pemex
Petróleos de Venezuela S.A.
Ecopetrol
Petrobras

Peto
Stato
Rio Tinto
Sh
BP
R
Total
Gler
Repsol
Sonatrach
Nigerian National Petroleum C

All of the companies shown here are natural gas and oil producers or mining companies. We know that burning coal is the form of energy production that's most harmful to the climate. Yet the energy industry insists

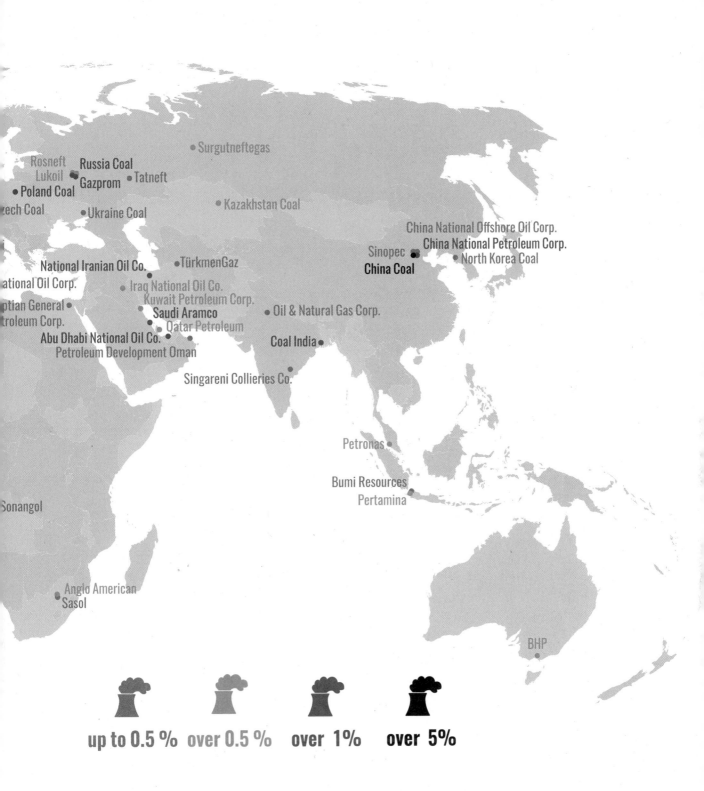

on coal remaining a major energy source for decades to come. All in all, three-quarters of coal used to generate electricity is burned in the Asian Pacific.

Note: on this map, some state-owned companies are represented collectively (China Coal, Poland Coal).

Countries with recycling rates of over 50 per cent

For a long time, waste in Europe was simply disposed of rather than recycled. In 2015, the European Commission agreed upon a new plan of action, which redefined waste as a potential raw material within the circular economy. Waste is now supposed to be reused and put back into the production chain so as to reduce demand for natural resources. Nonetheless, at present, in many countries more waste is still burned and dumped rather than recycled.

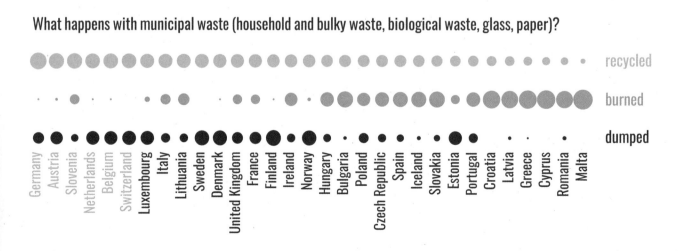

What happens with municipal waste (household and bulky waste, biological waste, glass, paper)?

no data

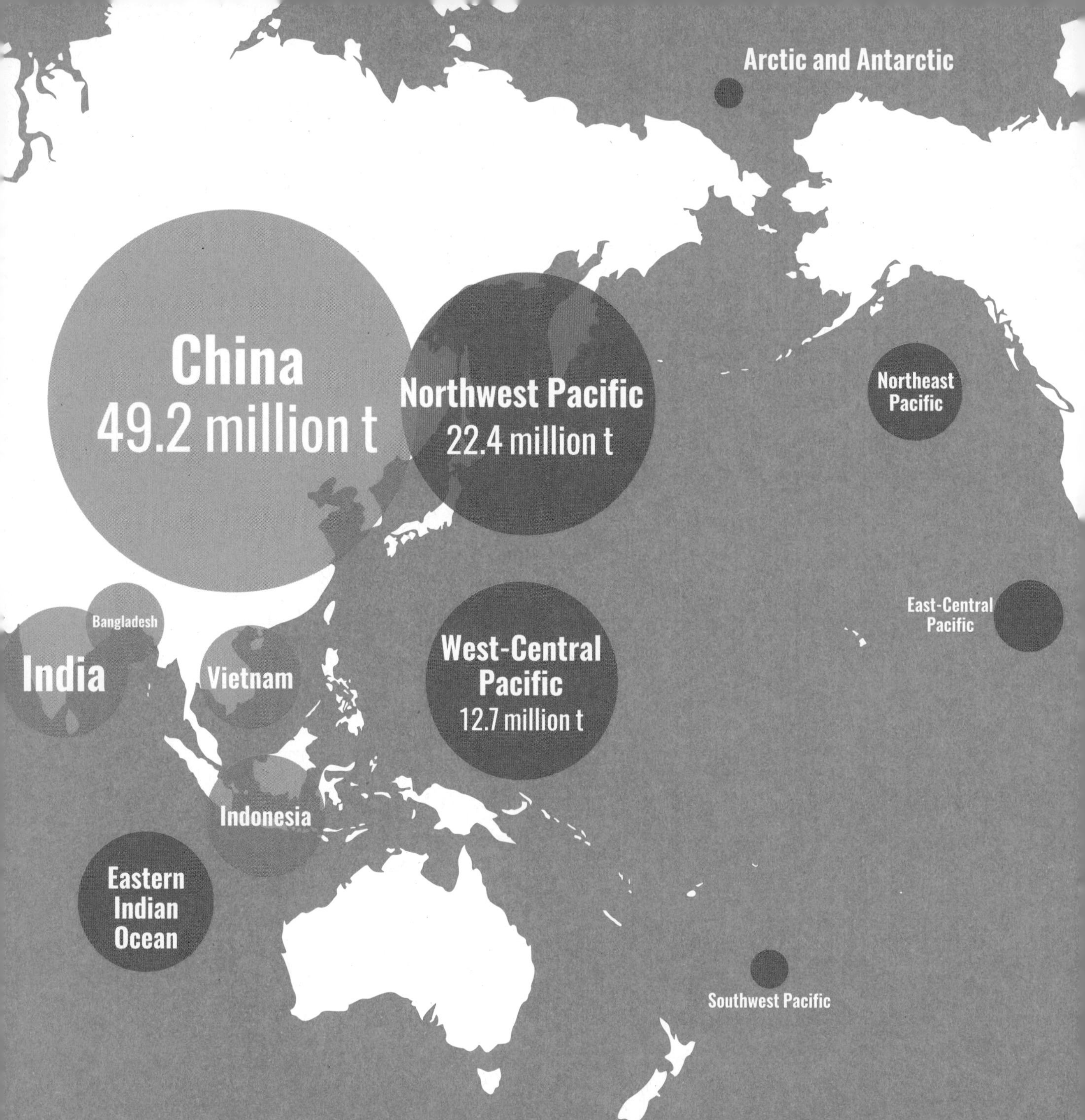

Arctic and Antarctic

China
49.2 million t

Northwest Pacific
22.4 million t

Northeast Pacific

Bangladesh

India

Vietnam

East-Central Pacific

West-Central Pacific
12.7 million t

Indonesia

Eastern Indian Ocean

Southwest Pacific

Where does our fish come from?

Don't believe everything it says on the label. What's advertised as 'Alaska salmon' may just be pollack, a poor cousin of cod, and it may not have been caught wild in Alaska but

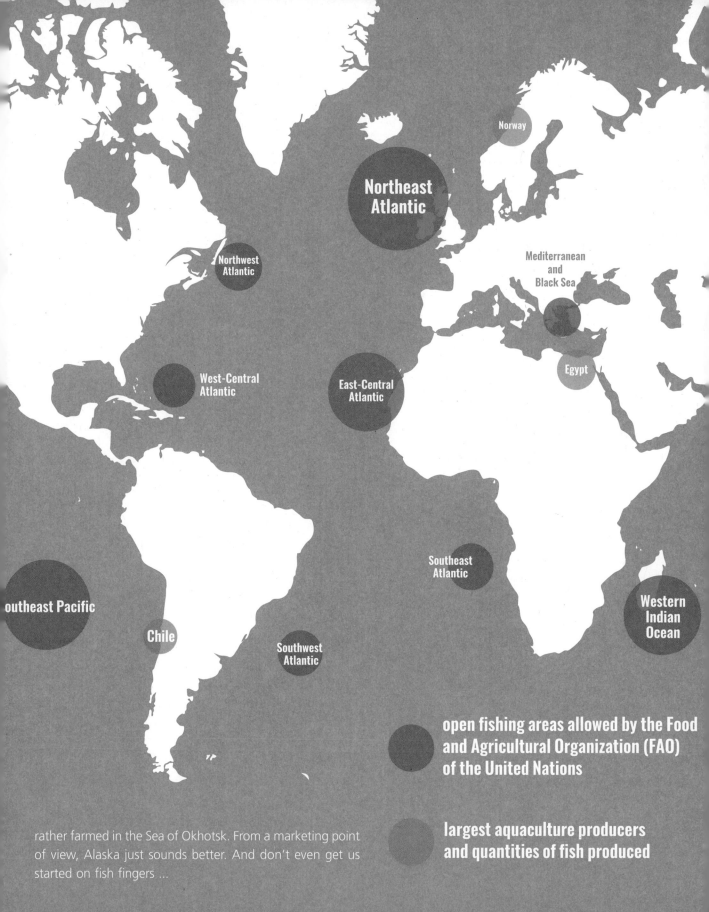

What makes 8 per cent of men see green?

Red

Only 13 per cent of the world's oceans are untouched areas of nature

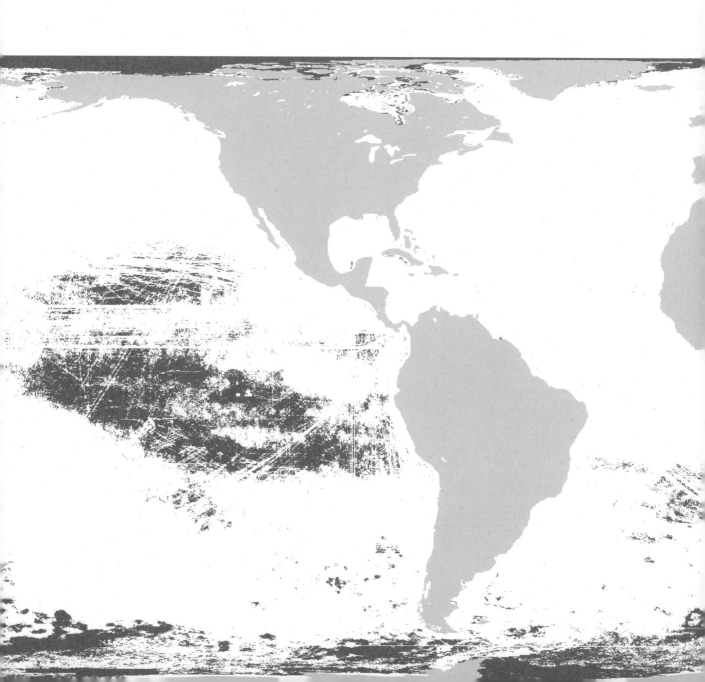

Coral reefs are disappearing, the oceans are over-fished and whole species are dying out. Where are the world's oceans still intact in their pristine condition? Researchers have found that the oceans have remained pure only in those places where there are few or no people – on the high seas and the Arctic and Antarctic. Only 5 per cent of oceans are natural conservation areas.

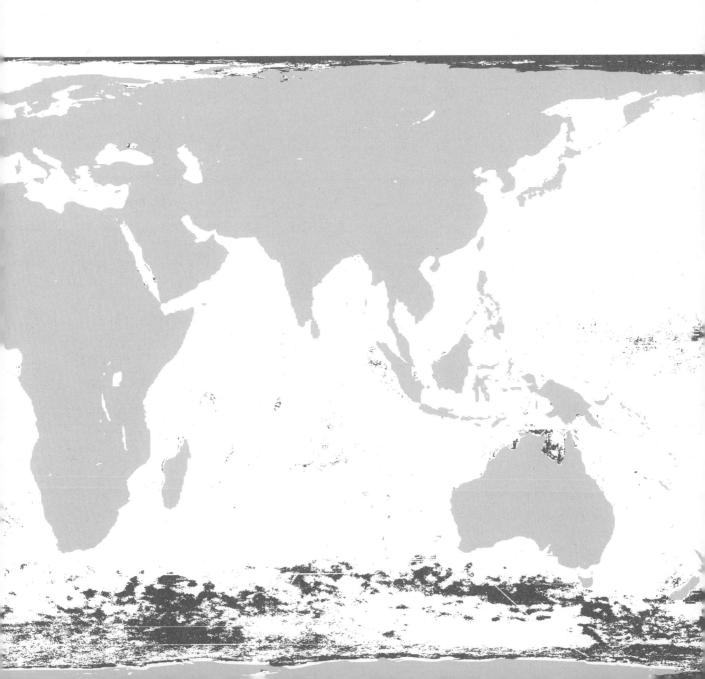

Number of cruise-ship dockings in selected ports in 2019

Tourism is a major source of income in many places, but masses of visitors bring a host of problems. For that reason, several port cities have been forced to develop strategies to preserve their historic centres and prevent them from being smothered in the smog generated by massive cruise ships.

In thinly settled Norway, for example, the port of Geirangerfjord is one of the most popular attractions in the country. Although only 200 people live in the nearby village of Geiranger, between 800,000 and 1,000,000 tourists visit it every year between May and September. That's the equivalent of 30 tourists per resident – every day. The situation becomes particularly problematic when several cruise ships anchor in the port at the same time. In 2018, some 200 of the floating behemoths arrived in Geirangerfjord.

In March 2019, the Norwegian government therefore dramatically tightened environmental regulations on ships entering Geiranger- and Nærøyfjord. It plans to continue to reduce the amount of emissions allowed step by step so that fewer ships will be permitted to dock.

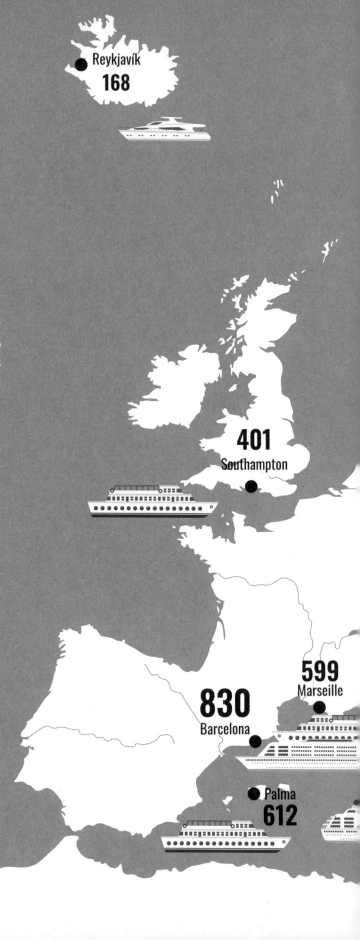

Reykjavík 168
Southampton 401
Barcelona 830
Marseille 599
Palma 612

The natural distribution of the coconut tree

The coconut tree owes its popularity to the fruit it bears. Ironically, coconuts are not nuts at all – they're actually some of the largest seeds in the world. The other remarkable thing about them is that they can float in salt water for long periods of time without losing their capacity to germinate. Botanists call the ability of seeds to travel across water hydrochoria. This capacity allowed the coconut – with some help from humans – to spread to tropical regions across the planet. Today coconuts are farmed intensively for their meat and water in Indonesia, the Philippines and India. Their shells can also be used as fuel for fires.

native coconut trees (eastern Africa, Sunda Islands)

neophytic (introduced from the outside) coconut trees (tropical coasts around the world)

The world's deadliest animals

Average number of humans killed per year, 2000–2010

crocodiles
• 1,000 victims

tapeworm

hippopotamus
• 500 victims

snakes

50,000 victims

dogs

40,000 victims of rabies

3.

freshwater snails

110,000 victims of transmitted schistosomiasis

2.

Homo sapiens

475,000 murder vict only

sharks — 10 victims

lions — 100 victims

wolves — 10 victims

elephants — 100 victims

2,000 victims

tsetse flies

 9,000 victims of transmitted sleeping sickness

kissing bugs — 12,000 victims of transmitted Chagas disease

roundworms

60,000 victims

1.

mosquitoes

725,000 victims of transmitted yellow fever, dengue fever and malaria

Where do the greatest number of bird species have their nesting grounds?

White-tailed sea eagles love places with little agriculture and few people. There are more and more of them in northern Europe because these birds of prey live on scraps left behind by feeding humpback whales in Norway or by swiping the kills of wolves in Finland. In Germany's forests too, there are now more than 700 nesting grounds for sea eagles.

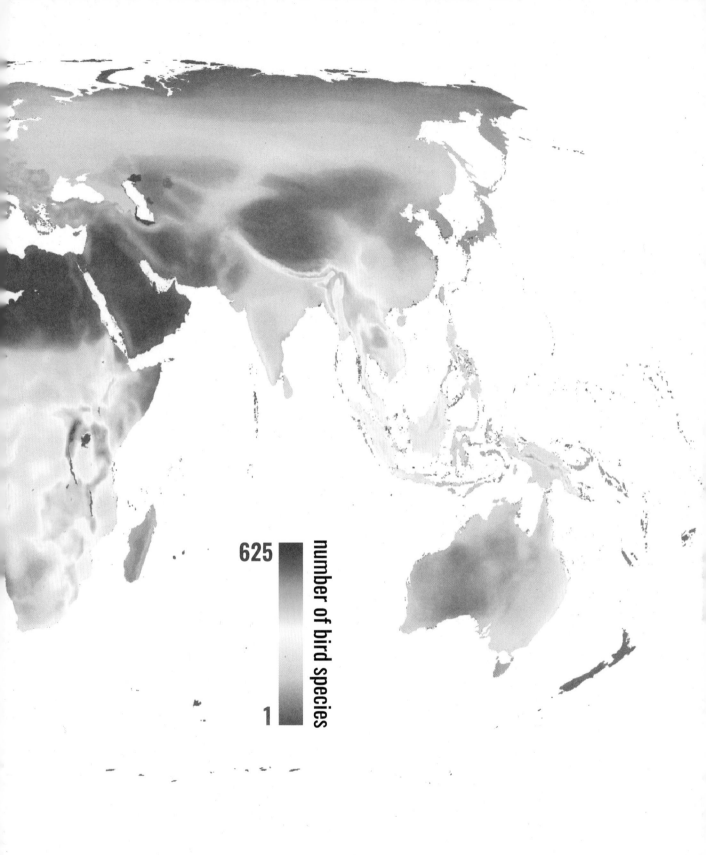

Things we're losing

Things we hate losing

Buying shoes online produces less CO_2 than buying them on the high street

If everyone did all their shopping online, it would not only be much quieter in city centres. It might even have a positive effect on the climate. The reason is that bricks-and-mortar shops need lots of energy and thus have negative CO_2 ratings. After all, the goods have to be stored and the lights have to stay on. And then there's the shopper's journey into town by car, bus or train.

Ordering things online from the comfort of your own sofa requires only making up your mind (0 grams CO_2) and a bit of electricity to power your computer (60 grams CO_2). There's one caveat, though: shoes and clothing are returned up to 80 per cent of the time, causing 370 grams of CO_2 not emitted on the high street. But in total, online shopping still has lower emission rates than going to the shops.

Travelling to a shop by public transport

20-km trip

1,710 g

trans-shipment and delivery 270 g
energy use of shop 1000 g
public transport 440 g

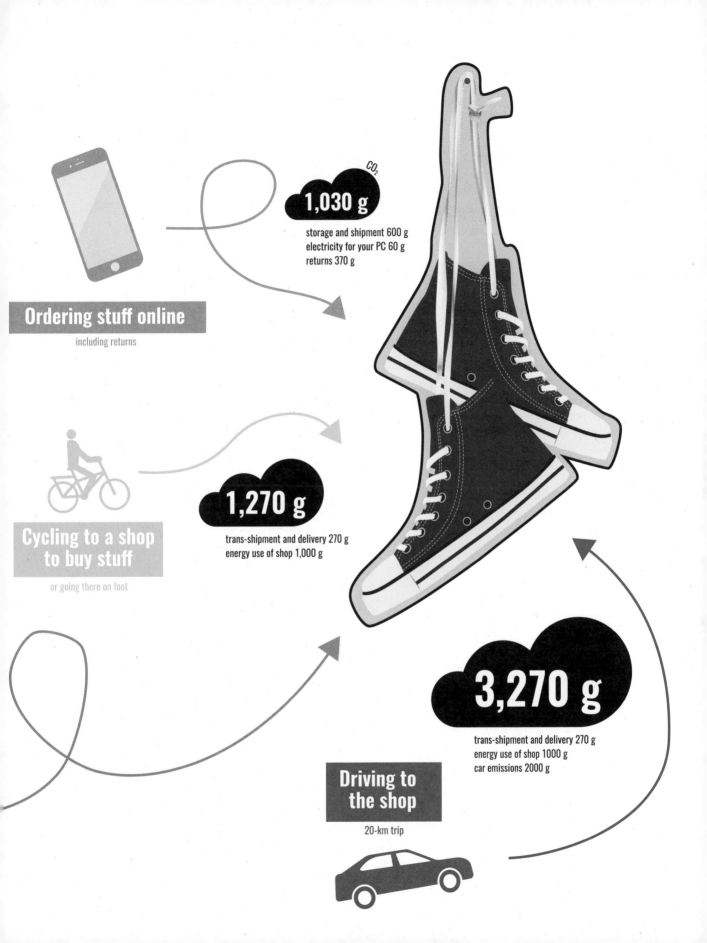

Maldives in 2019

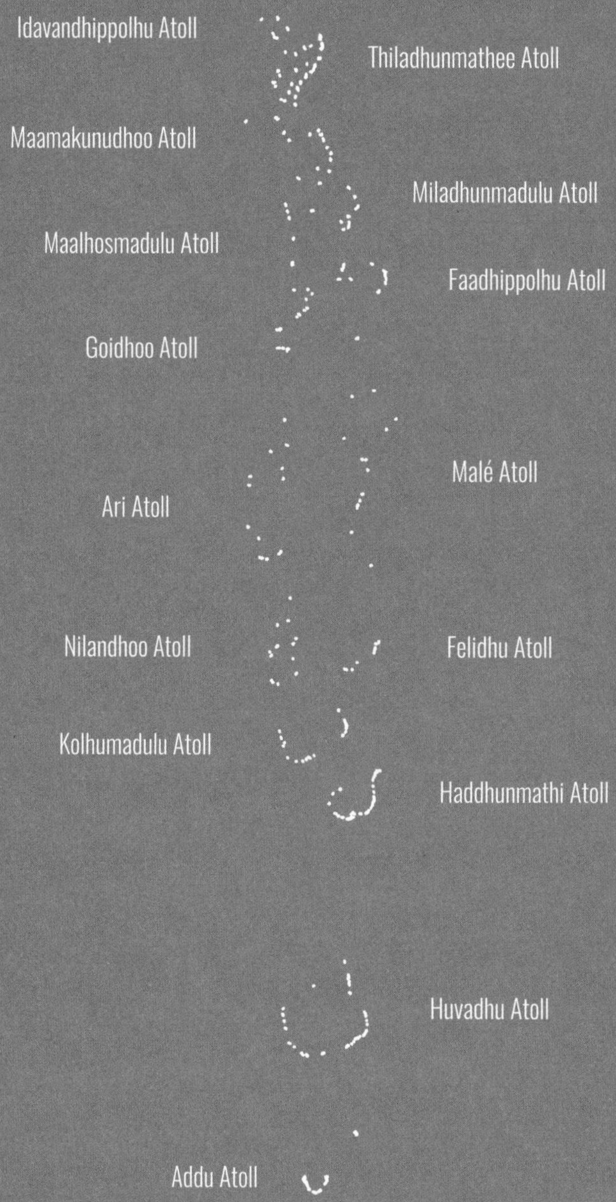

Idavandhippolhu Atoll

Thiladhunmathee Atoll

Maamakunudhoo Atoll

Miladhunmadulu Atoll

Maalhosmadulu Atoll

Faadhippolhu Atoll

Goidhoo Atoll

Ari Atoll

Malé Atoll

Nilandhoo Atoll

Felidhu Atoll

Kolhumadulu Atoll

Haddhunmathi Atoll

Huvadhu Atoll

Addu Atoll

Maldives in 2100

 highest points of the former Addu Atoll

If everybody in the world moved to Berlin, the city would be this big

Berlin is one of the hippest cities in the world. Today it has a population of around 3.7 million – but what would happen if everyone on earth decided to move to the German capital? If population density (4,090 people per square kilometre) remained the same, the city's square area would be increased by a factor of 2,113 from 891 to 1,882,640 square kilometres.

The furthest place from land on earth

There are at least 235 newspaper articles about Point Nemo, and they all start with something like this: 'Are you tired of people? Point Nemo is the most remote place on earth.' That's nonsense of course. If you want to be alone, you don't need to go as far as Point Nemo. But this location is interesting for another reason: space junk. The Americans, Chinese and Russians use it as a rubbish dump for obsolete satellites and the like – 260 spaceships have been dumped there since 1971. Here the sinking of space refuse is unlikely to bother anyone.

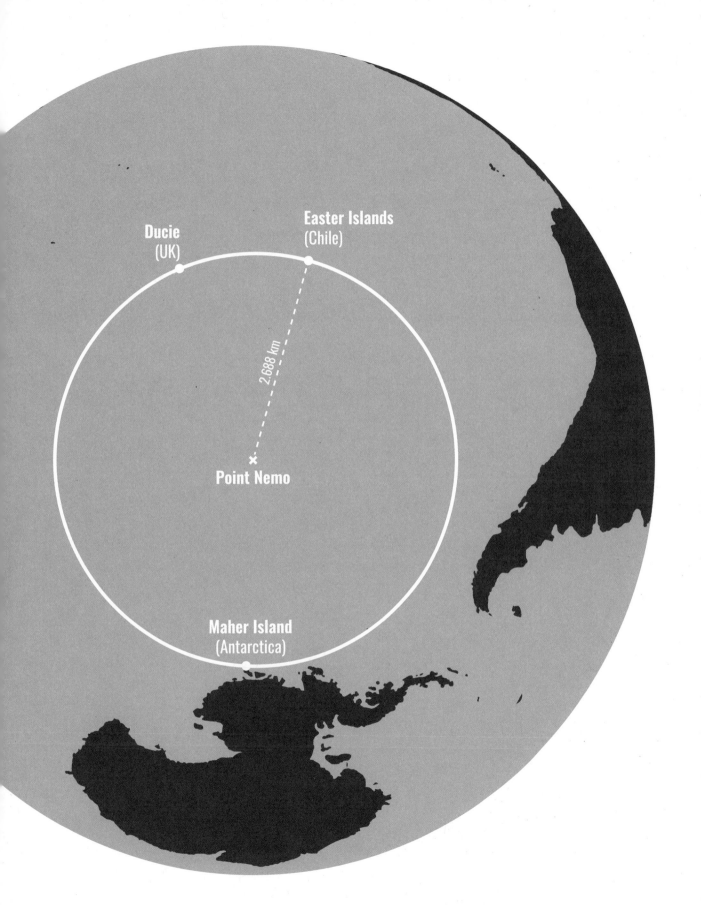

The CO_2 emissions caused by producing 40 copies of this or any other book is the equivalent of producing one e-book reader

Is using your e-reader ecologically better than collecting dusty novels on shelves? Well, sort of ... On the one hand, the printing and distribution of books cause environmental damage through cutting down trees for paper and emitting CO_2. But producing an electronic reading device also takes its toll on the environment. The CO_2 balance only tips in the e-reader's favour once more than 40 books have been read on the device.

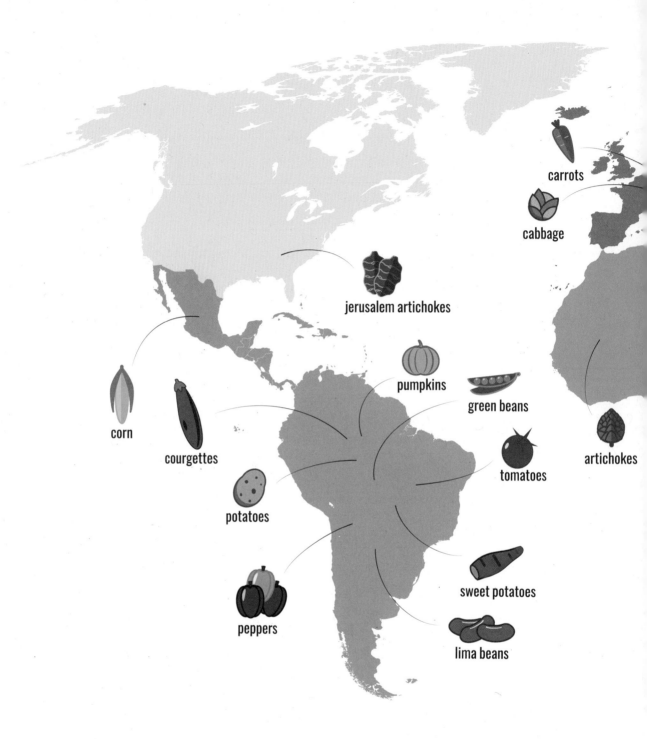

Where did our vegetables originally come from?

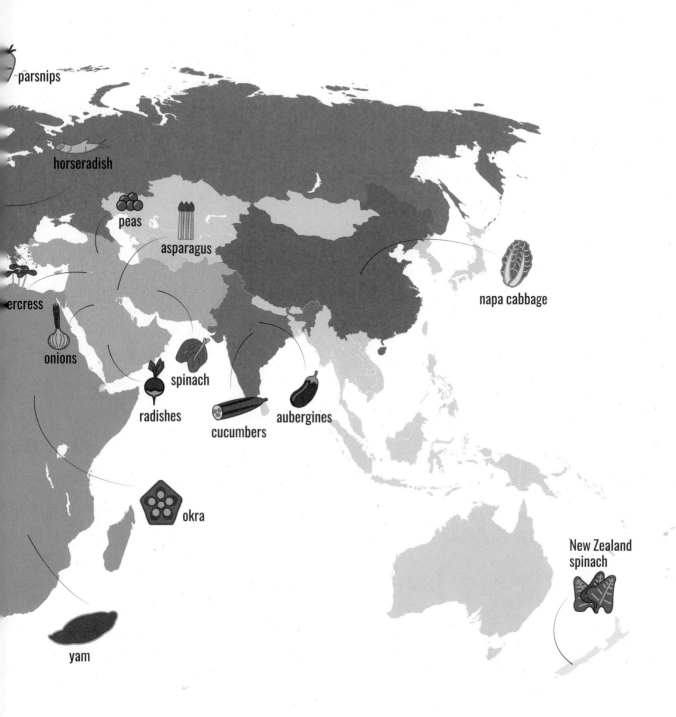

The tomato is the rock star among vegetables – or is it fruits? In any case, it first became a global hit in 1875 when it was made into ketchup. But it wasn't always like that. In previous centuries the tomato had been scorned after a handful of English and northern European aristocrats died after eating them. That was unfair, however. The real cause of death was the lead in the pewter plates used back then. The tomatoes' acidity extracted the lead from the plate – with fatal consequences for the aristocratic diners.

How many litres of milk does a cow produce
per day in the US?

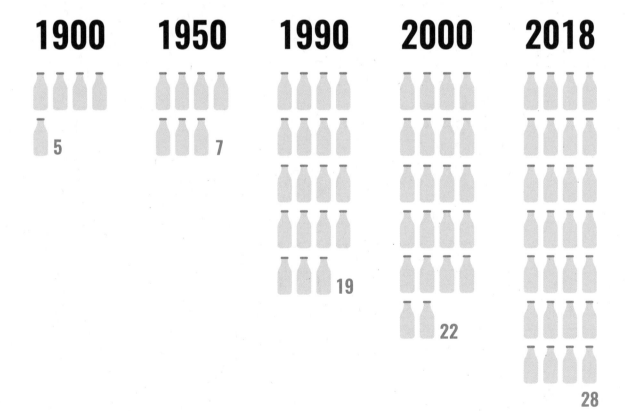

1900	1950	1990	2000	2018
5	7	19	22	28

Milk production

A cow today takes in roughly 50,000 calories a day, compared with only 10,000 a hundred years ago. That's because the cow is one of the most highly bred domesticated animals. Today's cows have huge, swollen and overburdened udders that can cause the animals considerable pain. The milk cow's natural life expectancy is twenty years. Today's industrial milk cows, by contrast, only live to the age of five.

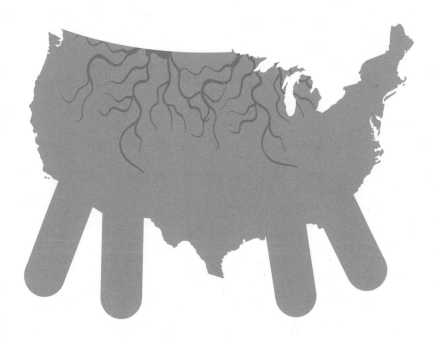

Nearly half of all environmental crimes in Italy in 2018 were committed in four regions

The regions in question have seen a boom not only in the business of illegal disposal of poisonous waste and other industrial contaminants, but also in the smuggling of animals and construction without permits. The so-called eco-mafia, which is organised into more than 350 families, earned 16.6 billion euros in 2018. The largest number of the known 28,000 ecological crimes in Italy were committed in Campania (3,862), Calabria (3,240), Apulia (2,854) and Sicily (2,641). There mafia families dispose of waste from all over Europe and sink whole ships full of toxic rubbish. Contaminated water and fields, together with incinerated industrial byproducts, may be responsible for an increase in cancer in these areas.

Investment in rail infrastructure in Europe

in euros per capita, 2018

Experts predict that rail travel will increase significantly in the coming decades, since compared to planes and cars, trains are the most environmentally friendly way of getting around.

With demand for tickets rising, many rail companies have introduced overnight trains and have recently been expanding high-speed connections. Europe's first high-speed connection was constructed in 1987 between Florence and Rome. It featured a train that could travel 250 kilometres an hour. In the meantime, this connection has been modernised and can now handle trains with speeds of up to 300 kilometres an hour.

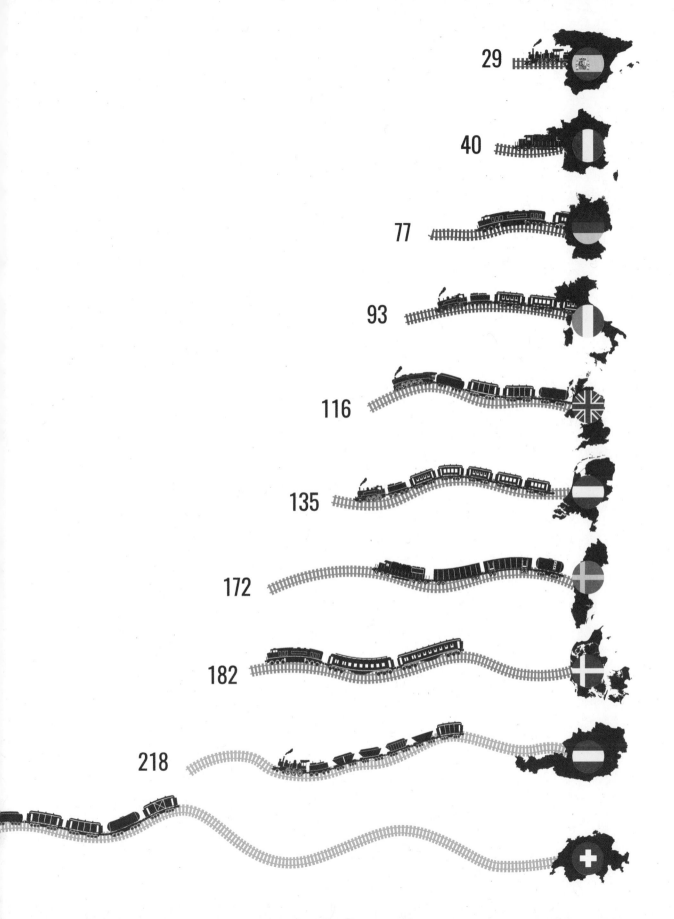

Mexico 274

Italy 190

France 145

Thailand 274

USA 160

Germany 144

Top 10 bottled water drinkers around the world
litres per head in 2018

Italians are big fans of bottled water, drinking more of the stuff than any people on earth except for Thais and Mexicans. But why? One litre of mineral water is 400 times more expensive than H_2O from the tap. What's more, tap water is subject to far stricter quality controls than bottled water, being examined for some 200 toxins whereas mineral water is checked for only 48. Yet despite all this, bottled water consumption in Italy has trebled in the past twenty years. Many Italians believe that mineral water is healthier, but that's not true. In fact, it's bad for the environment and dramatically increases water usage since water is needed to produce the bottles in which it's sold. So do the right thing and drink more tap water.

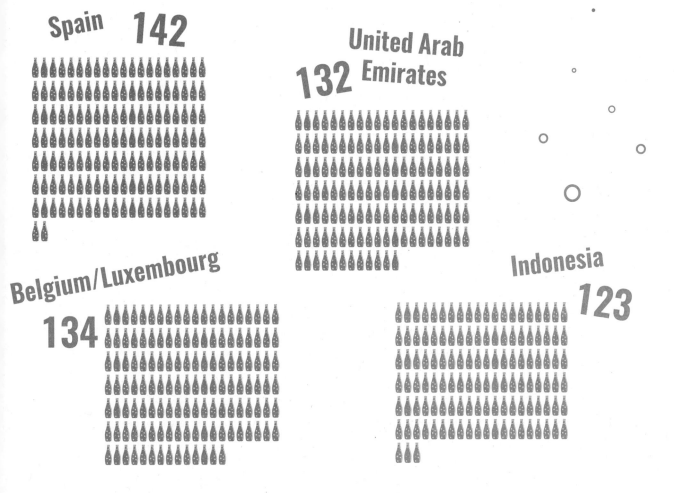

Find the dead fish

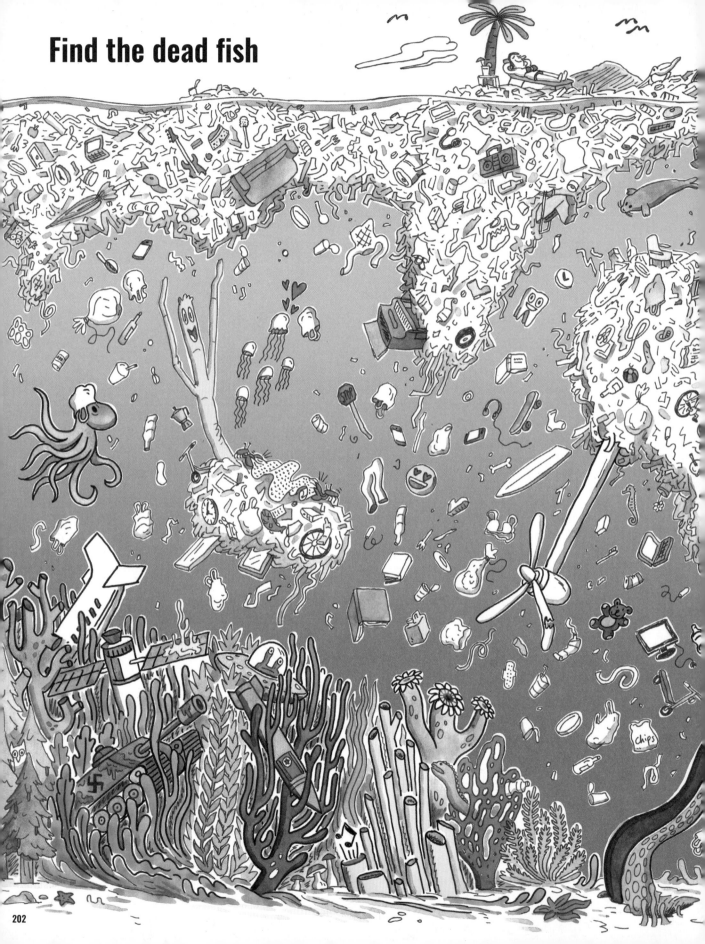

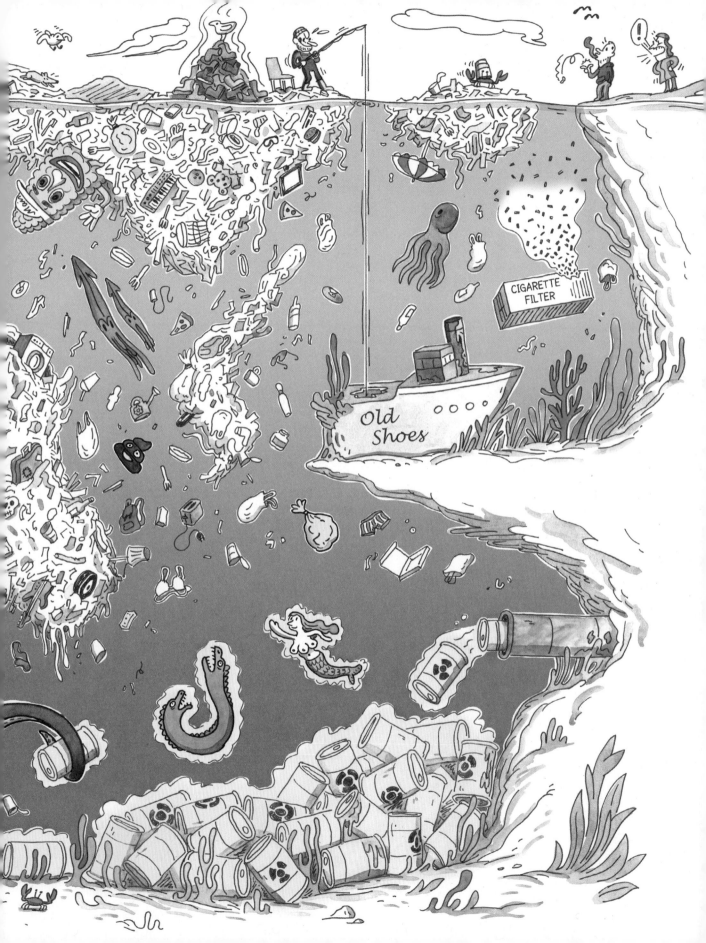

Sources

Active coal-burning power plants
Coal Swarm Project

Shark vs man in 2018
Floridamuseum.ufl.edu

Wolves in Europe
Europa.eu; wolf-mv.de

How much forest do we need to plant to make the world carbon-neutral?
Jean-François Bastin et al., 'The global tree restoration potential', *American Association for the Advancement of Science*, 365, no. 6448 (2019), 76–9; own projection

Reforestation is possible in the green areas
Ibd.

Every third piece of rubbish in the sea is a cigarette butt
Cleanupnetwork.com

Shrinkage of Switzerland's major glaciers between 1850 and 2010
Glacier Monitoring Switzerland; data.opendataportal.at; *Tagesanzeiger*

If all the people in the world stood next to one another …
Own calculations

Since human beings became sedentary 12,000 years ago …
Own calculations

Keeping a horse has the same impact on the environment as a 23,500-kilometre road trip
Jasmin Annaheim, Niels Jungbluth and Christoph Meili, *Ökobilanz von Haus- und Heimtieren: Überarbeiteter und ergänzter Bericht*, Schaffhausen, 2019

The Great Green Wall
Greatgreenwall.org

The highest and lowest points on Earth and Mars
Gunter Faure and Teresa Mensing, *Introduction to Planetary Science: The Geological Perspective*, Berlin, 2007; own research

Countries that index GNP; countries that index people's wellbeing
Wellbeingeconomy.org

Numbers of cruise passengers keep going up
Telegraph.co.uk; cruisemarketwatch.com; own research

Annual expenditure per person for cycle paths in Europe
Greenpeace; OpenStreetMap

Percentage of forest in national territories
World Bank

Minimalist maps
Naturalearthdata.com; OpenStreetMap; own research

Particulate matter and air quality in 2017
Earthobservatory.nasa.gov

The spread of raccoons and climate change
Vivien Louppe, Anthony Herrel et al., 'Current and future climatic regions favourable for a globally introduced wild carnivore, the raccoon Procyon lotor', *Scientific Reports*, 9, no. 9174 (2019)

Consumption of resources vs nutritional yield of various animals
Florian Fiebelkorn, 'Entomophagie – Insekten als Nahrungsmittel der Zukunft', *Biologie in unserer Zeit*, 2 (2017), 104–110; Florian Fiebelkorn and Miriam Kuckuck, 'Immer mehr Menschen mit Hunger auf Fleisch', *Geographische Rundschau*, 6 (2019), 48–51; own research

The five largest hydroelectric facilities …
US Energy Information Administration; ourworldindata.org; informationisbeautiful.net; Unipro; IAEA; IEA (data for 2018)

States that used a higher or lower percentage of renewable energies than the UK in 2018
Eurostat

A quick and easy way to stop smoking
Mathieu Vidard, *Science To Go: Merkwürdiges aus der Welt der Wissenschaft*, Munich, 2018, p. 41

Consumption of substances harmful to the ozone layer
UN Environment Programme Ozone Secretariat; ourworldindata.org

The forests of Europe
Globalforestwatch.org; Earthenginepartners.appspot.com

Trees as national symbols – official and unofficial
James Minahan, *The Complete Guide to National Symbols and Emblems*, Cirencester, 2009; own research

How much space do we need to supply the whole world with solar and wind power?
Desertec.org; Nadine May, *Eco-balance of a Solar Electricity Transmission from North Africa to Europe*, Braunschweig, 2005; own research

European countries that produce fewer greenhouse emissions per head than the UK
Eurostat

Natural catastrophes in 2018 and numbers of people affected
Internal-displacement.org

How much tropical forest does the earth lose every year?
Global Forest Watch; Agrar-Atlas 2019; regenwald-schuetzen.org

How much space per person do different countries require ...
Global Footprint Network

Forest growth in Europe
R. Fuchs et al., 'A high-resolution and harmonized model approach for reconstructing and analysing historic long changes in Europe', *Biogeosciences*, 10, no. 3 (2013), 1543–59; HILDA-Data for historical change in land use, University of Wageningen

How much progress have various countries made towards fulfilling the Paris Agreement?
Climate Action Tracker

What percentage of electricity was produced in Europe by burning coal in 2018?
Iea.org; beyond-coal.eu; own research

European cities renamed as those cities whose current temperatures they will experience by 2050
Jean-François Bastin et al., 'Understanding climate change from a global analysis of city analogues', *PLOS ONE*, 14, no. 7 (2019)

Global energy consumption between 1800 and 2017
Ourworldindata.org

Access to electricity in 2017
World Bank; IEA

The world's largest island of rubbish ...
Boell.de; Plastikatlas 2019

How much have temperatures risen in European cities since 1960?
Berkeleyearth.lbl.gov

Habitats of African elephants
WWF

Most common source of electricity in per cent (2017–18)
International Energy Agency; GoCompare Ltd; own research

Countries that produce fewer CO_2 emissions than those caused by global Internet porn consumption
The Shift Project

How many earths would we need if everyone lived like people in ...
DW.com

Annual CO_2 emissions in millions of tonnes by country
Global Carbon Atlas 2017; own research

Annual CO_2 emissions in millions of tonnes per capita
Global Carbon Atlas 2017; own research

Numbers of actively used mobile phones and toothbrushes worldwide
Theatlas.com

London to New York by car
CNN.com

How much oft the earth's surface has been paved over?
Globallandcover.com

CO2 emissions per litre of bottled water and tap water
GUTcert

Fish removed and rubbish added
FAO; German Federal Office for the Environment

Desertec plans for the distribution of renawable energy
Desertec.org; own research

Earthquake risk zones and selected earthquakes with high casualties since 1900
Natural Hazards Database on Earthquakes UNEP/GRID

Portland's path to becoming a bicycle city
ADFC; Ralph Buehler, *Street-Design fürs Fahrrad: Lernen vom Newcomer USA?*, Blacksburg, 2018; Portland Bureau of Transportation

Percentages of male and female smokers
World Health Organization

Human beings account for just 0.01 per cent of the earth's total biomass ...
Yinon M. Bar-On, Rob Phillips and Ron Milo, 'The biomass distribution on Earth', *PNAS*, 115, no. 25, (2018)

The ten largest seas by surface area
Own research

Annual toilet paper consumption per bottom in kilos
Statista

The ten worst maritime oil catastrophes
Marineinsight.com; own research

All the world's nuclear power plants
OpenStreetMap; IAEA; own research

Earth Overshoot Day comes earlier every year
Overshootday.org

Fires visible from satellites in the summer of 2019
Earthdata.nasa.gov (Active Fire Data)

Countries renamed as other nations with similar CO_2 emissions per capita
Global Carbon Atlas

70 per cent of global lithium deposits can be found in the orange triangle
US Geological Survey; own research

Countries which Germany criticises for failing to dispose of plastic waste properly …
Statista; own research

How much space would be needed to meet the global fuel demand with biofuel made of micro-algae?
Charles H. Greene et al., 'Marine microalgae: climate, energy, and food security from the sea', *Oceanography*, 29, no. 4 (2016), 10–15

Carpet of algae in the Atlantic gets bigger and bigger
Marine.usf.edu

Particulate matter, manmade
Earthobservatory.nasa.gov

These 67 companies emitted 67 per cent of all industrial greenhouse gases between 1988 and 2015
The Carbon Majors Database 2017; own research

Countries with a recycling rate of over 50 per cent
Sylvia Lehmann and Thomas Obermeier, 'Recyclingquoten – Wo stehen Deutschland, Österreich und die Schweiz mit dem neuen Rechenverfahren im Blick auf die EU-Ziele?', in Bernd Friedrich et al. (eds), *Recycling und Rohstoffe*, vol. 12, Neuruppin, 2019, pp. 85–98

Where does our fish come from?
Food and Agriculture Organization of the United Nations (ed.), *The State of World Fisheries and Aquacultures*, 2018

Only 13 per cent of the world's oceans are untouched areas of nature
Kendall R. Jones et al., 'The location and protection status of earth's diminishing marine wilderness', *Current Biology*, 28, no. 15 (2018), 2506–12.

Number of cruise-ship dockings in selected ports in 2019
Crew-center.com; own research

The natural distribution of the coconut tree
Biologische Invasionen: exhibition in the botanical gardens of the University of Potsdam

The world's deadliest animals
Floridamuseum.ufl.edu; own research

Where do the greatest number of bird species have their nesting grounds?
Clinton N. Jenkins, Stuart L. Pimm and Lucas N. Joppa, 'Global patterns of terrestrial vertebrate diversity and conservation', *PNAS*, 110, no. 28 (2013); biodiversitymapping.org; Birdlife International

Buying shoes online produces less CO_2 …
Oeko.de

Maldives in 2019 und 2100
Jonathan L. Bamber et al., 'Ice sheet contributions to future sea-level rise from structured expert judgment', *PNAS*, 116, no. 23 (2019); own research

If everybody in the world moved to Berlin, the city would be this big
Worldometers.info; Population Reference Bureau; statistik-berlin-brandenburg.de; own research

The CO_2 emissions caused by producing 40 copies of this or any other book …
Harish Jeswani and Adisa Azapagic, 'Is e-reading environmentally more sustainable than conventional reading?', *Clean Technologies and Environmental Policy*, 17, no. 3 (2014), 803–809.

Where did our vegetables originally come from?
Ciat.cgiar.org.

How many litres of milk does a cow produce in the US?
Normand St-Pierre and Michael Vandehaar, 'Major advances in nutrition: relevance to the sustainability of the dairy industry', *Journal of Dairy Science*, no. 4 (2006); United States Department of Agriculture

Nearly half of all environmental crimes in Italy in 2018 were committed in four regions
legambiente.it; spiegel.de

Investment in rail infrastructure in Europe
Allianz pro Schiene, on the basis of BMVI, VöV, MNVIT, SCI Verkehr GmbH

Top 10 bottled water drinkers around the world
Bottled Water Reporter, 49, no. 4 (2019); trekking.it

This book was put together by the team at KATAPULT:

Philipp Bauer, René Bocksch, Jonathan Dehn, Tim Ehlers, Benjamin Fredrich, Iris Fredrich, Lutz Fredrich, Julius Gabele, Sebastian Haupt, Christian Hildebrandt, Juli Katz, Nathanael Keidel, Christina Klammer, Jan-Niklas Kniewel, Felix Lange, Eva Pasch, Leonard Riegel, Ole Rockrohr, Cornelia Schimek, Stefanie Schuldt, Robin Siebert and Andrew Timmins.